Felix Publishing 2017
www.felixpublishing.com.au
email: info@felixpublishing.com
Print copies available from publisher.

Changing the Surface
Part of the Series **Adventures in Earth Science**
Other books in the series include:

    Exploration Science (Field Geology and Mapping)
    Riches from the Earth (Earth Materials and Mining)
    Rocks – Building the Earth
    Fossils – Life in the Rocks
    A Dangerous Planet (Earth Hazards)
    Through Sea and Sky (Oceanography and Meteorology)
    Beyond Planet Earth (Astronomy)

2016 digital book release
ISBN: 978-0-9946432-9-2
2017 Print Edition
ISBN: 978-0-9946433-0-8
Author: Dr P.T.Scott
All illustrations, photographs and videos by the author unless stated.
Cover photo: Wind eroded pedestal rock, the Arbol de Peidra or Stone Tree, near Uyani, western Bolivia (Photo: Matthew Scott), Design after AJS Creative, Brisbane.

Registration:
Thorpe-Bowker +61 3 8517 8342
email: bowkerlink@thorpe.com.au
No part of this publication may be reproduced, stored in a retrieval system, or transmitted in any form or by any means, electronic, mechanical, photocopying, recording or otherwise, without the prior written permission of the publisher.

© All rights reserved Felix Publishing

CHANGING

THE

SURFACE

Dr. Peter T. Scott

First released 2017
All rights reserved Felix Publishing

To my grandchildren who are
yet to find their own adventures.

About the Author

Dr. Peter Scott is an award-winning teacher of Earth Science of over forty years' experience in both Secondary and Tertiary Education. He holds a Bachelors' Degree, two Masters' Degrees and a Doctorate including many years on his own research in locating and correlating coal measures. He has visited many places of interest including Antarctica, the Andes, the Amazon, North Africa, volcanic islands of the Pacific and Asia, California northern Europe and remote Australia.

Dr. Scott at Tinajari, the Devil's Canyon, near Lampa, Peru 2016

Table of Contents

Chapter 1: Weathering and Erosion	1
1.1 Introduction	1
1.2 Physical Weathering	11
1.3 Chemical Weathering	19
1.4 Types of Chemical Weathering	26
1.5 Soils	30
Chapter 2: The Chemical Weathering of Carbonate Rocks	43
2.1 Introduction	43
2.2 Karst Topography	43
2.3 Limestone Caves	46
Chapter 3: The Work of Moving Water	53
3.1 Introduction	53
3.2 River (Fluvial) Erosion and Deposition	54
3.3 Coastal Erosion and Deposition	66
Chapter 4: The Work of Moving Ice	78
4.1 Introduction	78
4.2 Alpine Glaciation	79
4.3 Continental Glaciation	93
Chapter 5: The Work of Moving Wind	100
5.1 Introduction	100
5.2 Major Features formed by Wind	101
5.3 Major Features Formed by Wind and Water	107
Summary	113
Practical Tips	116
Multichoice Questions	118
Review and Discussion Questions	124
Answers to Multichoice Questions	126
Reading List	127
Key terms Index	129

Chapter 1: Weathering and Erosion

1.1 Introduction

As soon as new igneous rock material is exposed to the surface of the earth due to uplift and erosion, the minerals at the exposed faces of the fresh rock are subjected to the chemical and physical effects of the weather. Minerals are chemical compounds and apart from gold, silver and native copper, they react with other chemicals in their new the environment.

The breakdown of these new minerals, and therefore the rocks which contain them, lead to the formation of new secondary minerals, or the simple physical powdering of more resistant minerals such as quartz. These new products are the raw material which can form soils and, under other conditions, new rock types. These in turn can be broken down anew and thus become part of a great re-cycling process of Earth materials.

For example, in the igneous rock granite, which is formed deep underground but is often exposed at the surface following tectonic uplift, the individual minerals become less defined as their crystal faces become more rounded and weathered. Black and dark-coloured ferromagnesian minerals containing much iron and magnesium such as hornblende and biotite, usually weather to red-brown iron oxides such as haematite and limonite. Pink and white feldspars, such as orthoclase and plagioclase, weather to clays and soluble minerals. The light grey quartz does not weather, but is broken into smaller pieces and masked by the iron oxide and clay. Overall, the weathered granite is more rounded and friable, that is, it is able to be easily crumbled.

Figure 1.1: The weathering of granite.

Once a rock has been made more friable, and some of its minerals converted to softer, more soluble forms or reduced to small particles by **weathering**, it is then able to be abraded and moved elsewhere by the **erosion** due to gravity, wind, water and ice. Weathering is the alteration of rock by:

- chemical attack by water, gases in the air and natural acids and solutions in the soil

- temperature changes

- action of plants and bacteria

This attack occurs where the rock has been exposed to these agents of weathering (in situ). Almost all weathering takes place in the zone between the Earth's surface and that of the water table, the top level of water within the Earth's surface layers. In this zone, the physical and chemical conditions are suited to the rapid breakdown of material. After a time, erosion will lower the surface of the land, as well as the water table, exposing fresh rocks to the weathering processes.

Water is a major part of both weathering as a chemical agent, and an eroding medium as a carrier of abrading particles and medium of transport. The water of the Earth in its various forms above and below ground is recycled as part of a great **water cycle** (or hydrological cycle).

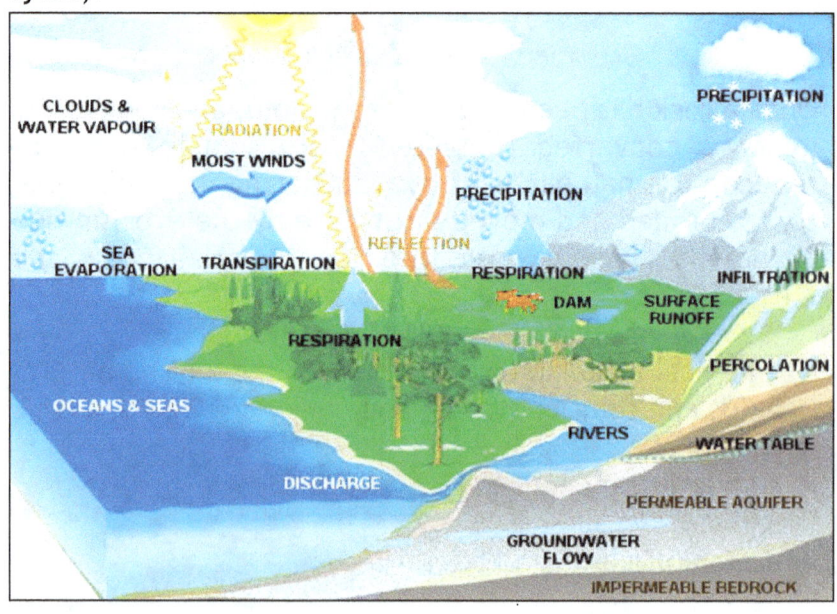

Figure 1.2: The hydrological cycle showing the recycling of the world's water

Erosion is a more general term describing the processes whereby rock is loosened or dissolved and then removed elsewhere. It often includes the processes of weathering but also those of dissolution, abrasion and transportation. The main agents of abrasion and transportation which produce major landscapes are:

- gravity
- water (from rain, rivers, sea)
- ice (due to frost, glaciers)
- wind

Any process, including the four given above, which moves rocks under the influence of gravity, is generally termed **mass wasting**. Erosion removes rock and sediment which has previously been formed by weathering. To do this, the rock must be uplifted by forces within the Earth, and exposed at the Earth's surface where it can be acted upon by the agents of erosion. The amount of erosion depends upon the:

- nature of the rock as some rocks erode more than others

- climatic conditions such as wind, rain, temperature changes

- slope of the land, which is the angle of the land surface to the horizontal. The minimum angle at which a slope will retain loose material is called the

angle of repose. Some of these angles for different sediments are given below:

MATERIAL	ANGLE of REPOSE
Ash	40°
Clay (dry)	25-40°
Clay (wet)	15°
Dirt (dry)	30-45°
Gravel	25-30°
Sand (wet)	45°
Sand (dry)	34°
Sand (water filled)	15-30°
Snow	38°

Table 1.1: The angles of repose for some materials. Notice that clay and sand which is filled with water have their angle of repose greatly reduced, i.e. these sediments will begin to slide at relatively gentle slopes

Gravitation plays a major part in the reshaping of the Earth's surface. Any material which is able to flow or fall will be carried down from high places to low by the pull of the Earth's gravity. This will occur with loose, solid material which has been broken by chemical and physical weathering, as well as within masses of moving water and ice. Some of the major forms of mass wasting by gravity alone include:

- **Slumping** is the general term used for the downslope slipping of a mass of soil or rock.

- **Soil Creep** is the very slow movement of soil downhill. It is more pronounced in areas where the soil has a high clay component such as in black soil regions, where the soil has been formed from basalt and similar rock. Here, fence posts, telegraph poles and even trees, are tilted at angles as their soil base slowly creeps downslope. The steeper slopes may looked rippled and folded into many small, parallel ripples.

Figure 1.3: Soil creep on a hillside overlooking Auckland, New Zealand

- **Solifluction** is also a gradual mass wasting slope process but this is caused by the constant freeze-thaw processes which occur in cold climates. **Frost heave**, or the lifting of the rock and debris by ice as it freezes and expands, causes the material to loosen. This is then moved further down slope when the ice thaws.

Figure 1.4: Solifluction on a slope in Switzerland (Photo: Wikipedia)

- **Talus slopes** are steep slopes of broken, angular rock material called **scree** which forms at the base of a cliff or mountain as gravity causes the loose material to roll downhill.

Figure 1.5: A Talus slope in the Andes, southeast of Santiago, Chile

- **Landslides** are the more dramatic and sudden movement of rock and debris downhill. These can be triggered by excessive weight, especially from the building of man-made structures, earth tremors or by the addition of water during heavy rains.

Figure 1.6: a dramatic landslide on a de-forested hillside in Mexico (Photo: Wikipedia)

- **Mudflows** consist of water-saturated soil which flows very rapidly down a slope and often follows heavy rainfall, especially in regions where there has been deforestation. A **lahar** is a special form of fast-flowing mudflow formed when heavy rains or snow and ice melt on the unstable slopes of ash volcanoes. This is often triggered by new volcanic activity which will cause the melting of snow covering the upper slopes. Many volcanoes of the high Andes are subject to this form of mass-wasting, often with disastrous results. On November 13, 1985, a small eruption of the volcano Nevado del Ruiz, produced an enormous lahar which buried the town of Armero in Colombia, causing an estimated 25,000 deaths.

Figure 1.7: A huge lahar coming from Mount St. Helens in Washington State, USA after the 1982 eruption (Photo: USGS)

Figure 1.8: Destruction in the town of Armero, Colombia, after the inundation of the lahar of the 1985 eruption of the Nevado del Ruiz volcano (Photo: USGS)

- **Avalanches** are masses of snow, mixed with air which suddenly move down a mountain under the influence of gravity. They often come from the sudden collapse of large, overhanging masses of snow called **cornices**. These cornices, form on the upper slopes in alpine regions and are a common hazard in these places. They can be triggered by a sudden thaw during warmer weather, by additional loads imposed by skiers, animals or explosives or from local seismic activity. In alpine communities, avalanche fences are often built upon potentially unstable snow slopes and covers are built over roadways and rail lines.

Figure 1.9: A large avalanche in the Rocky Mountains (Photo: US National Parks Service)

- **Turbidity currents** are underwater avalanches of debris which suddenly break off the coastal continental shelf and then flow rapidly down slope. These form the submarine canyons often found along the coastlines of some continents and they form major deposits of sediments called **turbidites** on the ocean floor.

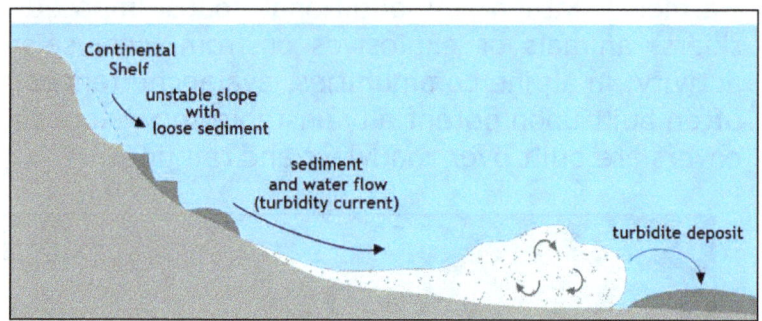

Figure 1.10: Diagram showing a turbidity current

In addition to these natural processes, considerable erosion also takes place because of the action of animals and humankind. There is also a group of rocks and soil which contain clays, called **thixotropic clays** which are generally hard and adhesive until shaken. When vibrated by earth tremors or man-made construction, they suddenly liquefy and collapse.

1.2 Physical Weathering

This is the disintegration of rock by physical means by which the rock simply breaks down with little or no change in its chemistry. This can be caused by:

- **Exfoliation** (or onion skin weathering) occurs due to the repeated cycles of hot and cold temperatures which cause differential expansion and contraction in hard crystalline rocks such as granite, gabbro and similar rocks. The outer layer of usually only a few millimetres will peel off, producing a rounded surface.

Figure 1.11: Diagram showing exfoliation (at left) and the process on basalt (right), Antrim, Northern Ireland

- **Off-loading** occurs on a larger scale when deep-formed crystalline rocks such as granite crack into large, curved sheets as the rock is uplifted and the massive weight of the overlying rock is taken off by erosion. The removal of this rock, and its applied pressure, causes the outer surfaces of the crystalline rocks to crack because of their internal forces which once opposed the overlying pressure.

Figure 1.12: Slabs of granite peeling off due to off-loading, south-western Tasmania, Australia

- **Frost wedging** occurs when water freezes, and expands as it changes to crystalline ice. Water freezing in very small fissures within the rock will expand and enlarge them, eventually breaking the rock apart. This action occurs mainly in cold, mountain regions and is responsible for the sharp, angular nature of alpine landscapes.

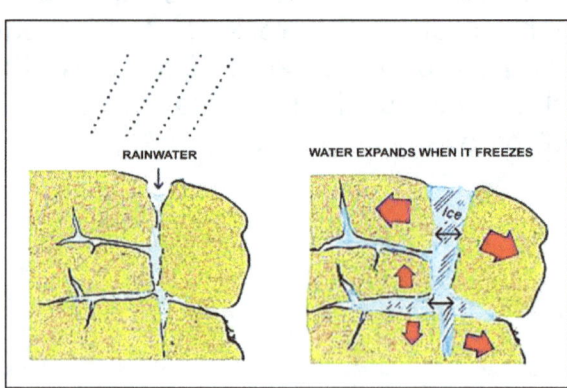

Figure 1.13: Diagram showing frost wedging

- **Hydration** is the addition of water to a mineral's crystal structure which causes their expansion e.g. anhydrite ($CaSO_4$) expands as it absorbs water to became gypsum ($CaSO_4 \cdot 2H_2O$). Clay minerals also have this effect which helps to expand and break up rock and soils contain clays.

- **Abrasion** is the wearing-away of rock and soil due to the grinding effects of fragments carried by downhill by gravitation, by water, by wind and by ice. Glacial and some wind-formed (aeolian) rock fragments are angular and sharp, whereas those rocks abraded by water are usually rounded in shape.

- **Fault brecciation** occurs within faults, causing the rock along the fault line to be broken rock down to flour size as the fault moves and the surfaces are ground together. The weakness in the rock due to the fault filled with permeable material can then allow for further erosion by water and other agents.

Figure 1.14: Fault breccia in the plane of a normal fault

- **Salt wedging** occurs on cliffs and rock which is subject to regular contact with salty sea spray, especially where the rock is porous such as sandstone. Saltwater spray is thrown up onto the rock surface where the water component evaporates. As it does so, minute salt crystals form and wedge out the small grains of the rock at its surface. This produces a pock-mark effect with rounded hollows varying from a few millimetres to several metres in depth.

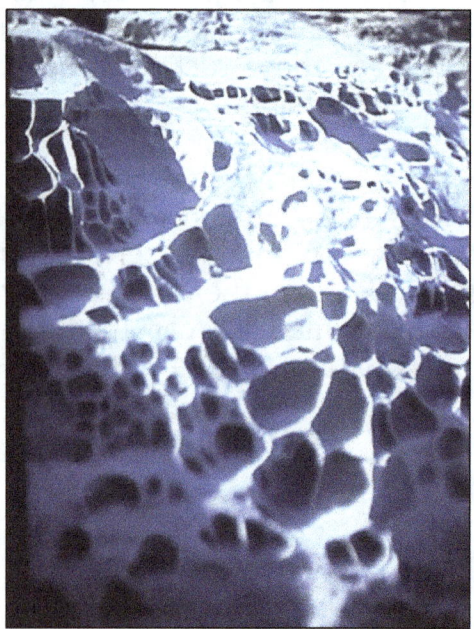

Figure 1.15: Salt pitting in the coastal sandstone cliffs near Maroubra Beach, Sydney, Australia

- **Biological activity** can also play a major role in the physical break up of rock. Lichen and other rock-penetrating organisms can eventually break down a rock surface. Larger plant roots are capable of widening minute cracks and penetrating parent rock.

Figure 1.16: Orange lichen (*Xanthoria elegans*) breaking down a rock stack, Half-Moon Inlet, Antarctica.

Figure 1.17: Tree roots cracking rock, Mt. Cootha, Brisbane, Australia

One of the main indicators that there is some large igneous rock structure, such as a batholith just below the surface, is the presence of large rounded boulders or **tors**. These are often found as a huge complex of balancing rocks with bare rocky surfaces. These are formed by a combination of off-loading and exfoliation as the hard, crystalline rock is brought near the surface by uplift and then exposed by erosion. These rocks crack due to the unloading of the vast amount of material removed by erosion and its subsequent expansion. The cracks are further enlarged by both physical weathering and chemical attack by water and atmospheric gases so that the edges of the rock along the cracks are slowly removed and rounded.

Figure 1.18: Jointed rock exposed (left), weathered along joints (centre), and then the eventual rounding of edges and removal of material to form tors

Figure 1.19: Granite tors near Cooma, NSW, Australia

Figure 1.20: A balancing tor near Warialda, NSW, Australia

1.3 Chemical Weathering

Rock surfaces exposed to the gases and solutions of nature soon chemically change and gradually decompose as their minerals individually react. Even man-made objects which use stone as a building material also suffer from chemical decomposition because of the elements of the weather and additional air pollution.

The gases in the air, oxygen (21% by volume), nitrogen (78%) and carbon dioxide (0.05%) directly or indirectly dissolved in water and may combine with the minerals in rock to produce softer, more soluble products which are then able to be more easily eroded by physical means.

Natural waters may contain dissolved gases producing mild acids, which, given sufficient time, will dissolve many of the natural minerals. Carbon dioxide gas readily dissolves in water to form weak carbonic acid:

$$CO_2 + H_2O \longrightarrow 2H^+ + CO_3^{3-}$$
$$\text{carbon dioxide gas} \quad \text{water} \quad\quad \text{carbonic acid}$$

This process is but one of the several reactions and exchange of carbon dioxide in nature as part of the great re-cycling system called the **carbon cycle**.

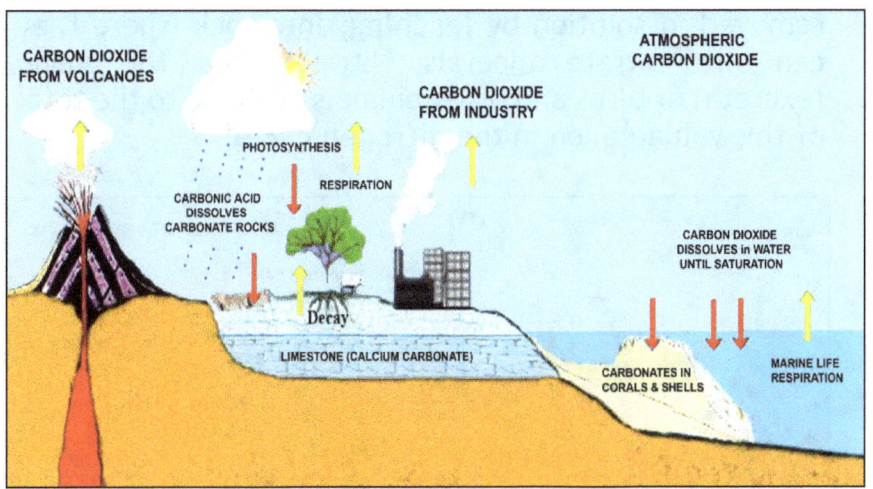

Figure 1.21: Diagram showing the carbon cycle

Nitrogen dioxide gas is also involved in mineral change. It is formed by nitrogen and oxygen gases in the air combining during lightning emissions and then dissolving in water. The nitric oxide gas (NO), then reacts with more free oxygen gas to produce more nitrogen dioxide (NO_2) for more reaction and the nitrogen dioxide dissolves in water to form nitric acid.

$$3NO_2 + H_2O \longrightarrow 2H^+ + 2NO_2^- + NO \uparrow$$

nitrogen dioxide gas   water   nitric acid   nitrous oxide gas

In the soil and within the nodules of plants, especially legumes such as beans, bacteria also fix nitrogen gas to other atoms to form more soluble ammonia (NH_3). The nitrates and ammonium ions thus produced are then taken up by plants to make protein. Nitrates are also removed in solution by **leaching**, into rock where they can form nitrate minerals. Nitrates from the guano (excreta) of birds and other animals also add to the total of this valuable ion in the **nitrogen cycle.**

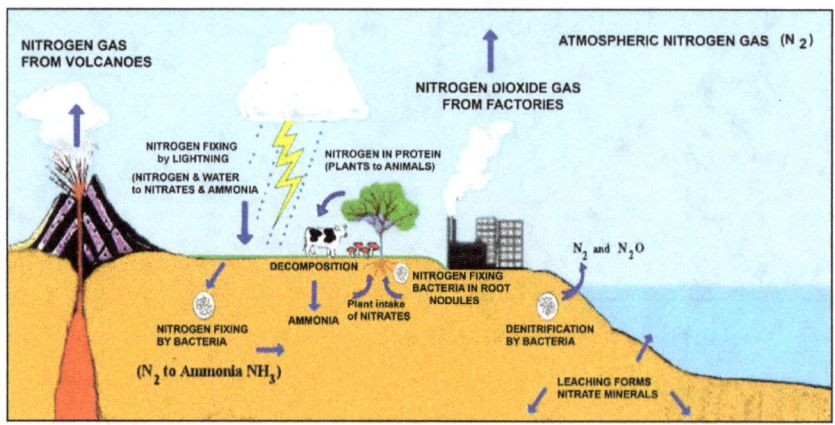

Figure 1.22: Diagram showing the re-cycling of nitrogen

Sulfur dioxide gas, especially from volcanic gases and industry, also readily dissolves in water to produce acids:

$$SO_2 + H_2O \longrightarrow 2H^+ + SO_3^{2-}$$
sulfur dioxide gas   water   sulphurous acid

Whilst gases such as nitrogen and carbon dioxide react with minerals in solutions, oxygen gas is reactive enough to directly produce oxides with most minerals when they are exposed to the air, at or near the surface.

The red and yellow oxides of iron such as haematite and limonite, for example, are very common on the Earth's surface in a variety of rocks and mineral deposits. This was not always the case. The Earth only achieved an oxygenated atmosphere after blue-green algae (cyanobacteria) began to produce oxygen gas by the process of **photosynthesis** and this process is still used by green plants today:

$$6CO_2 + 12H_2O \xrightarrow{\text{sunlight, chlorophyll}} C_6H_{12}O_6 + 6O_2 \uparrow + 6H_2O$$

carbon dioxide gas + water → glucose sugar + oxygen gas + water

The oxygenation of the atmosphere probably began about 2300 million years ago. Although cyanobacteria had been around much earlier than this, much of the early free oxygen gas produced by them would have reacted with minerals to form the many oxide minerals found today. Some of the earliest Pre-Cambrian sedimentary rocks contain considerable amounts of iron oxides in rock deposits. These deposits known as banded iron formations, show the development of an oxygenated atmosphere.

An important result of the oxygenation of the atmosphere was the formation of the **ozone** layer. This is a belt of gases, from 20 to 30 kilometres altitude, containing a higher concentration of ozone molecules (O_3) which absorb harmful ultraviolet (UV) radiation from the Sun. Ozone gas is chemically unstable and absorbs the UV radiation to produce normal oxygen (O_2):

$$O_3 \xrightarrow{\text{UV radiation}} O_2 + O^*$$

ozone → oxygen + unstable oxygen atom (free radical)

The unstable atomic oxygen quickly reacts with normal oxygen molecules to form more ozone:

$$O^* + O_2 \longrightarrow O_3$$

unstable oxygen atom + oxygen → ozone

More recently, concerns have been raised about the depletion of ozone due to some organic molecules. These man-made **organohalogen** compounds, especially chlorofluorocarbons (CFCs) and bromofluorocarbons (BFCs), which have been released into the atmosphere by human activity react with the ozone molecules. This breakdown occurs when a UV ray strikes a CFC or BFC molecule removing a single, unstable chlorine atom as a **free radical**, which is an atom with an unpaired electron which makes it very unstable and reactive. This free radical then combines with an ozone molecule, tearing away a single oxygen atom to form an unstable molecule of chlorine monoxide. This then combines with other free radical oxygen atoms forming atmospheric oxygen, leaving the chlorine atom free to destroy another ozone molecule:

1. $$CFCl_3 \xrightarrow{\text{UV radiation}} CFCl_2 + Cl^*$$

trichlorofluoromethane (a typical CFC) → dichlorofluoromethane + unstable chlorine atom (free radical)

2. $$Cl^* + O_3 \longrightarrow ClO + O_2$$

unstable chlorine atom (free radical) + ozone → chlorine monoxide + oxygen

As there are many free radicals of oxygen (O) present in the atmosphere because of the oxygen gas-ozone reaction (see above), the unstable chlorine monoxide reacts with them, producing more unstable free radicals of chlorine (Cl*):

$$ClO + O^* \longrightarrow Cl^* + O_2$$

chlorine monoxide   unstable oxygen atom (free radical)   unstable chlorine atom (free radical)   oxygen

The free chlorine atom is then free to attack another ozone molecule and so continue more breakdown of ozone (reaction 2).

Other dissolved gases, such as ammonia (NH_3) and hydrogen sulphide (H_2S), as well as natural mineral salts and organic products from decay, also provide reactive solutions able to slowly weather the surfaces of rocks.

In general, the weathering of rocks occurs at different rates and by different chemical processes with different minerals. Chemical weathering depends upon:

- Surface area of exposure which increases the area of contact for chemical and physical change

- Temperature affects the rates of chemical reaction usually increase with temperature. Chemical weathering seems to be more extensive in hotter climates

- Humidity or water content of the air which not only reacts with minerals, but also allows for solutions of gases and ions which then also react with minerals

- Vegetation cover which physically breaks up the rock and provides humic acids from the soil and more carbon dioxide.

- Mineral composition of the rock which weathers in reverse order to that of their formation from molten rock as given in Bowen's Reaction Series.

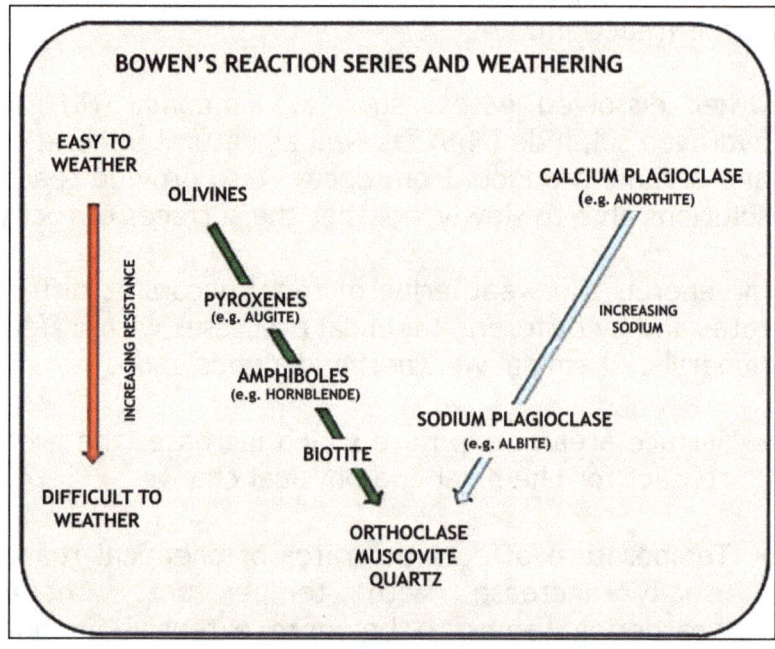

Figure 1.23: Bowen's Reaction Series applied to the rate of weathering

As a quick guide, the following is generally true for the most common rock-forming minerals:

- **Feldspars** ⟶ **Clay Minerals** ⟶ **Bauxite + Solutions**
  (aluminium silicates   (hydrated aluminium   (complex aluminium
  of potassium, sodium   silicates)   oxide-hydroxide)
  and calcium)

- **Micas** ⟶ **Clay Minerals + Iron Oxides + Solutions**
  (silicates with   (hydrated aluminium   (if iron was in
  aluminium,   silicates)   the mica)
  potassium, iron)

- **Ferromagnesian** ⟶ **Clay Minerals + Iron Oxides + Solutions**
  Minerals
  (iron & magnesium   (hydrated   (e.g. red haematite,
  silicates)   aluminium silicates)   yellow limonite)
  e.g. hornblende
  augite, biotite

- **Quartz** ⟶ No change but will physically erode into sand

1.4 Types of Chemical Weathering

Some of the most common types of chemical weathering include:

- **Oxidation** occurs when oxygen gas, a very reactive chemical element, from the air is dissolved in water, reacts with minerals in rocks to form oxides which are often softer and more easily eroded. Many of these, such as the oxides of iron - haematite (Fe_2O_3), limonite ($Fe_2O_3 \cdot nH_2O$) and magnetite (Fe_3O_4) - are valuable ores. For example, ferromagnesian minerals such as olivine.

$$Fe_2SiO_4 + O_2 + H_2O \rightarrow Fe_2O_3 + H_2SiO_4$$

forsterite   oxygen   water   haematite   silicic acid
(a form of   gas   (unstable)
olivine)

- **Hydration** and **hydrolysis** concern the interaction of the water molecule with minerals. Hydration is the absorption of water molecules by a mineral whereas hydrolysis involves the break up or dissociation of the water molecule into hydrogen ions (H^+) and hydroxyl ions (OH^-) which then react with mineral in the rock. Water (H_2O) is one of the most effective agents of weathering. The shape of the water molecule causes it to be an electric dipole, that is has a positive and a negative end, which enables it to react to the electric fields within the crystals of the mineral.

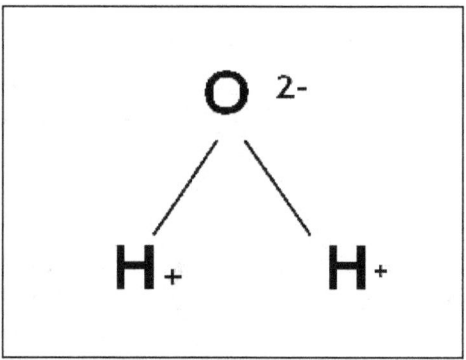

Figure 1.24: The water dipole

Consequently, water molecules weaken the electrostatic forces of attraction between ions within the crystal lattice. As well, the water molecules may group around certain ions, especially **cations** or ions of positive charge and hydrate them e.g.

$$CaSO_4 + 2H_2O \longrightarrow CaSO_4.2H_2O$$
anhydrite   water   gypsum

$$2Fe_2O_3 + 3H_2O \longrightarrow 2Fe_2O_3.3H_2O$$
haematite   water   limonite

Furthermore, water ionises ($2H_2O \rightarrow H_3O^+ + OH^-$), producing hydrogen ions (H^+ usually in the H_3O^+ form) and hydroxyl ions (OH^-) which also react with the other ions of mineral crystals. This is the process of hydrolysis. For example, in some silicate minerals, ions from absorbed water will enter the crystal structure and displace metal ions. The hydrogen ions H^+ or hydroxyl ions OH^- (present in water), are very small and thus have a concentrated electric field which easily displaces the metal ions.

$$3KAlSi_3O_8 + 2H^+ + 12H_2O \longrightarrow KAl_3Si_3O_{10}(OH)_2 + 2K^+ + 6Si(OH)_4$$
orthoclase   hydrogen   water   illite clay   potassium   soluble
                ions   ions   hydrated
                                                                                       silica

The mineral illite, under further hydration, can become kaolinite clay and then gibbsite which is part of bauxite, the complex oxide-hydroxide of aluminium - the main ore of this metal.

- **Solution** is when some minerals directly dissolve in water, forming solutions which may then be washed away.

$$NaCl \longrightarrow Na^+ + Cl^-$$
halite   sodium & chlorine
               ions in solution

At Hallein in Austria (from whence the terms halite halogens are derived), the Celts mined salt for thousands of years by channelling water into their tunnels in the subterranean salt dome, dissolving it and then running the solution out of the mine and into evaporation pans.

- **Carbonation** is when the gas carbon dioxide (from the air plant and soil processes) slowly dissolves in water to form weak carbonic acid.

$$2H_2O \quad + \quad CO_2 \quad \longrightarrow \quad 2H^+ \quad + \quad CO_3^{2-}$$
water   carbon dioxide gas   carbonic acid

Over time, this acid will attack and dissolve any carbonate mineral, such as calcite (calcium carbonate), magnesite (magnesium carbonate) and siderite (iron carbonate). Many monomineralic rocks, rocks containing mainly one mineral, are made up mainly of a carbonate mineral. For example, limestone (calcium carbonate), marble (a metamorphosed limestone), dolomite (calcium-magnesium carbonate) and marl (calcium carbonate rich mudstone). In addition, many other rocks have calcite as cement, holding grains of other minerals together in the rock as in calcareous sandstone. These carbonate-rich rocks are then dissolved slowly, usually under the water table, or broken up as their cement dissolves.

1.5 Soils

Pedology from Greek: pedon for soil and logos for study, is the study of soils in their natural environment. Soils can be considered as the end product of weathering and erosion which is capable of supporting plant life. It is chiefly composed of physically and chemically weathered rock and mineral products, as well as decaying vegetable matter called humus. Soils, broken rock material, dust and other loose material which covers the Earth's surface is generally called **regolith**. The type of soil present in a local area depends upon:

- Local rock type, because weathering occurs immediately where rock is exposed to the chemical and physical agents near or at the surface. The minerals within the rock are the basic components which then break down to form the soil. For example:

- Rocks such as quartz-rich sandstones, granites, rhyolites and some schists usually form light-coloured, sandy soils.

- Rocks such as basalt and gabbro which contain high amounts of ferromagnesian minerals such as biotite, hornblende and augite, will weather to rich red and brown soils reflecting the high iron content of the parent rock.

- Rocks with a high amount of feldspar will produce clay soils.

- Limestone and similar calcareous rocks will produce poor limey soils.

- Topography or the surface of the land, will affect the amount of soil which will accumulate on the slope of the land. Steep slopes will not have much soil cover as most soil will move down slope under gravity, to be deposited as sediment in lower basins on land or under the ocean.

- Erosion of soil by water, ice and wind will take it elsewhere, often long distances from where the original source rocks were weathered. A dramatic example of this is the formation of deep, rich soils of **loess**. This is very fine soil which is blown by wind off the surfaces of deserts and areas of glacial erosion, and then deposited over wide areas a considerable distance away.

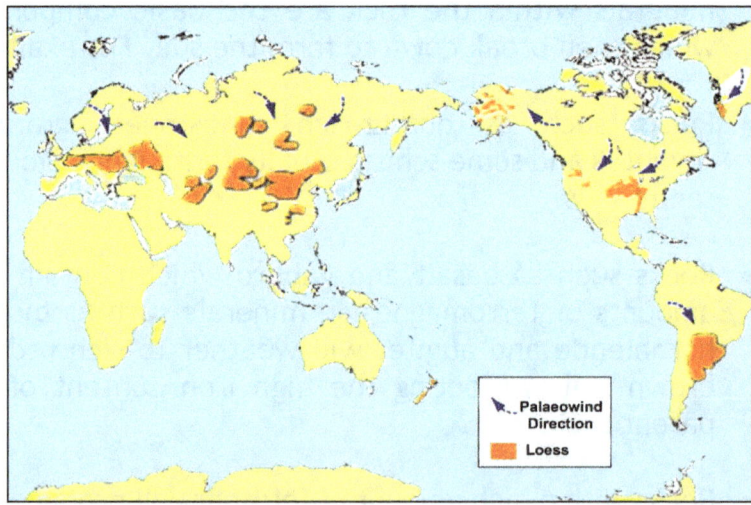

Figure 1.25: Worldwide loess distribution and possible ancient wind direction

- Climate will also determine soil types, as chemical and physical weathering depends upon the amount of heat and water in the local environment. Physical weathering depends upon changes in temperature to expand and contract the rock or freeze water within fine cracks. Chemical weathering is due to the chemical reactions of the minerals within the rock with water, atmospheric gases and other reactive compounds and the rate of such reactions also depends upon the local climate. In tropical regions, chemical weathering is more pronounced whereas physical weathering is more prevalent in cold and arid regions.

- Vegetation cover affects soils because the amount of vegetation locally will determine the amount of the dark, rich organic material, **humus** which can develop in the soil. This nutrient material will only form if there is sufficient temperature and moisture for good plant growth. In arid hot and cold areas where there is little vegetation, the soils are generally free of this rich material. Also, if the rainfall is too high, much of the rich nutrient in the soil is washed out, or leached, and so the soils then are also poor. When a soil is rich in humus, more plants will grow to renew the soil-vegetation cycle.

- Time will allow undisturbed rock material to be thoroughly weathered. In tropical regions and with rocks which weather quickly, soils can be formed in a relatively short space of time. For example, in the tropical Pacific region, Islanders will plant coconut palms in piles of recently-formed basaltic volcanic rock knowing that soil will develop quickly.

Other factors which should be considered when studying soils are the:

- Bulk of the soil, which consists of the main products of weathering such as pebbles, sand and clay. Soils having a high ratio of sand to clay will not hold water as well as very clayey soils. Loam is a good, fertile soil containing roughly equal proportions of sand, silt and clay with some humus.

- Mineral salts present as ions in the soil. These salts are derived from chemical weathering and are important for the production of vegetation. The most essential ions are:

Phosphate	PO_4^{3-}
Nitrate	NO_3^-
Sulfate	SO_4^{2-}
Iodide	I^-
Potassium	K^+
Iron II	Fe^{2+}
Calcium	Ca^{2+}
Magnesium	Mg^{2+}

  For a fertile soil, there are also some other ions which are needed as trace elements, including:

Copper II	Cu^{2+}
Cobalt	Co^{2+}
Chromium	Cr^{3+}
Manganese	Mn^{2+}
Nickel	Ni^{2+}

- Humus content which not only holds the grains of the topsoil together but also retains moisture. As soils mature, the humus content gradually increases.

- Living organisms within the soil which influence plant growth, such as the nitrogen-fixing bacteria which convert atmospheric nitrogen gas into soluble nitrate ions which can then be absorbed into the plant's roots. Soil is also aerated by larger organisms such as earthworms.

Usually soils are studied in cross-section as **soil profiles** which are a cross-sectional view through the soil at a particular location. It consists of layers, running parallel to the surface, called **soil horizons**. At any given location, the nature of the soil changes with depth, with the surface processes involving the atmosphere and possible plant interaction, changing to those involving the bare rock below. There are many variations in representing a soil profile and it will also vary with different soil types, however a simplified profile could be shown as:

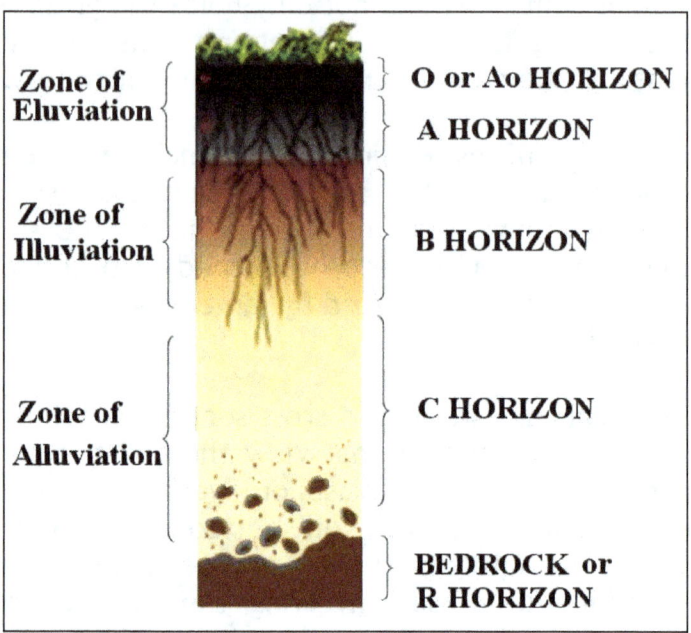

Figure 1.26: A diagram of a typical soil profile

The main horizons include the:

- O Horizon consists of the organic matter of leaf litter and humus. This can vary in size from non-existent in places of low vegetation cover, to well over a metre in tropical areas. In some classifications it is considered as part of the A horizon and designated as Ao.

- A Horizon (or topsoil) is the first horizon with a rock-derived mineral content and its composition will depend upon the type of rock from which the soil has been derived and the weathering and accumulation processes which has deposited it. Its colour will also depend upon these factors and the amount of organic matter mixed in from the horizon above. The O and A Horizons are in the zone of **eluviation**, that part of the soil which is penetrated downwards and laterally by water and dissolved or suspended material.

- B Horizon (or subsoil) is the where the products of weathering and soil reaction accumulate by leaching from above. It may be rich in clay from weathered feldspars, in calcium minerals derived from carbonate rocks such as limestones, or in red-yellow iron oxides, such as haematite, from weathered ferromagnesian minerals as found in basalt, gabbro and similar rocks. It also may contain a considerable amount of aluminium oxides. This horizon is a zone of **illuviation**, that part of the soil which receives solutions and suspensions which have percolated down from above such as clay.

- C Horizon is a place with few transformations, chemical change and with only some movement of soluble products and further oxidation of iron. Processes here also depend upon the thickness of the horizons above. This horizon is composed of the fragments derived from the bedrock below and is sometimes denoted as an R Horizon or the zone of **alluviation**. It may contain particles ranging in size from small grains to large boulders and often a mixture of both.

Depending upon the thickness and complexity of the soil, some horizons may be further sub-divided into many transitional zones. In some places, such as desert pavements there may be little or no soil.

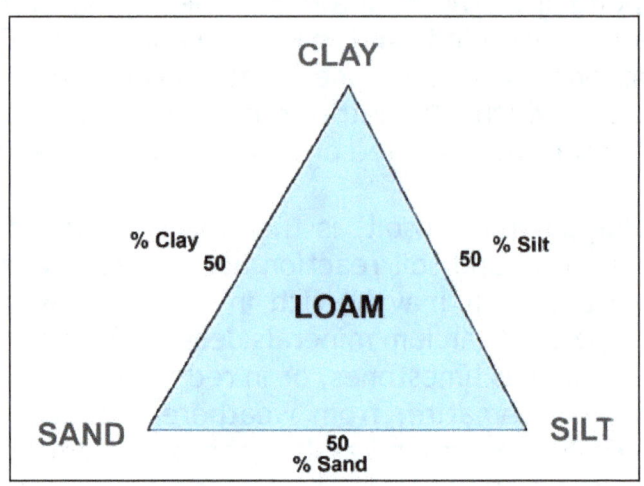

Figure 1.27: Diagram showing a triangular graph for soils

There are many different classifications of soil depending upon country of location and the practical application for the use of the soil. Engineers are more likely to be concerned with the physical characteristics of soil whereas agronomists (soil scientists), would be more concerned with the application of the soil for plant and crop production. A very basic classification uses the texture of the soil, described by three endpoints of sand, silt and clay, with a middle zone containing all three elements combined as loam. This can be represented in a triangular graph such as the one shown in Figure 1.27. These graphs can also be used to show transition soils such as sandy loams, silt clays, clayey loams, and so on, with their location on the graph depending upon their percentages for each of the endpoint element along the sides of the graph. Other, more scientific classification systems for identifying and naming of soils vary greatly. One classification system of soil classification is that of the Natural Resources Conservation Service (NRCS) of the Department of Agriculture of the United States. This is shown in Table 1.2 on the next page. For easy identification these characteristics are shown in Figure 1.28 below.

Figure 1.28: Diagram showing a pictorial soil classification of the Natural Resources Conservation Service (NRCS) of the US Department of Agriculture

NAME	MAIN FEATURES
Alfisols	Soils which form under forest vegetation where the parent material has undergone significant weathering, being light in colour in surface horizons, illuviation of clay in the B horizon and moderate to high concentrations of potassium (K^+), calcium (Ca^{2+}) and magnesium (Mg^{2+}) ions further down.
Andisols	Soils which develop from volcanic parent materials and have an accumulation of allophane (an amorphous hydrous aluminium silicate clay) and oxides of iron and aluminium in the developing soil.
Aridisols	Soils that develop in very dry environments with poor and shallow soil horizon development, light in colour (because of limited humus) and with the deposition of salts (carried upward by the soil water) at or near the ground surface because of evaporation. This soil process is of course called salinization.
Entisols	Soils which are immature and lack the vertical development of horizons. These soils are often associated with recently deposited sediments from wind, water, or ice erosion (such as loess) and have variable moisture content.
Gelisols	Soils of cold climates permanently frozen within 2 m. of the surface. A horizon of organic material rests on the frozen zone.
Histosols	Soils which form in areas of poor drainage, have thick accumulations of organic matter in their horizons and with variable moisture content.
Inceptisols	Soils which are immature but more developed than entisols and are often found in arctic tundra environments, glacial deposits, and relatively recent deposits of stream alluvium. They have an immature development of eluviation in the A horizon, illuviation in the B horizon and the beginning of weathering processes on parent material sediments.
Mollisols	Soils common to grassland environments (often now under cultivation) and have a dark-coloured A horizon (due to a rich humus), a rich base and are quite fertile. In more arid environments these soils often exhibit calcification with surface layers which are saturated with calcium solutions.
Oxisols	Soils which develop in tropical and subtropical regions with high rainfall and temperature. Profiles are featureless, containing mixtures of quartz, kaolin clay, iron and aluminium oxides and some organic matter. The abundance of iron and aluminium oxides found in these soils results from strong chemical weathering and considerable leaching and may contain laterite formed by a seasonal and fluctuating water table.

Spodosols	Soils which develop under conifer or pine forests and so are modified by podsolization which is the downward migration of aluminium and iron ions (Al^{3+}, Fe^{2+} and Fe^{3+}), together with organic matter, from the surface areas and their accumulation in the profile's deep areas. Parent materials of these soils tend to be rich in sand the migration of ions contributes to acid accumulations in the soil. They have little clay and only small quantities of humus in their A horizon.
Ultisols	Soils in regions of warm temperatures and abundant availability of moisture which enhances the weathering process and increases the rate of leaching. This also causes the dominance of iron and aluminium oxides with the presence of the iron oxides causing the A horizon to be stained red. They have low levels of quantities of K^+, Ca^{2+} and Mg^{2+} and moisture at base.
Vertisols	Soils heavy in clays which show significant expansion and contraction due to the presence or absence of moisture. They are common in areas that have shale parent material and heavy precipitation.

Table 1.2 Major soil classifications of the Natural Resources Conservation Service (NRCS) of the Department of Agriculture of the United States

Soils can be degraded and their fertility reduced by removal of the top soil by wind and other forms of erosion such as leaching and **salination**. Leaching is the loss of water-soluble plant nutrients from the soil, due to rain and excessive irrigation. This is often increased by deforestation as the removal of vegetation assists the water runoff and loss of nutrients as well as erosion in general. Salination is caused by the introduction of excessive salt (sodium chloride) to the soil. This can occur due to encroachment of seawater near the coast as the land slowly sinks or if irrigation aquifers are contaminated by seawater. It can also occur inland if too much irrigation is applied in arid regions. Here, the water table is raised by the additional water being added by the irrigation. As the water table rises, it may bring with it salt which has been removed from the rocks below which were originally formed with a considerable salt content (from a desert or marine habitat). A more recent form of salination is occurring in many of the low-lying Pacific islands, such as in Kiribati. Here, sea level rise due to global warming is bringing seawater up into the permeable limestones of these coral islands and destroying coconut plantations from below.

Figure 1.29: Diagram showing salination of a Pacific island coconut garden by an increase in height of the water table

Chapter 2: Chemical Weathering of Carbonate Rocks

2.1 Introduction

The major types of weathering and erosion produce characteristic features and landscapes due to their unique processes. In places where the rocks have high carbonate compositions such as limestone ($CaCO_3$), marble (metamorphic limestone) and dolomite ($Ca\,Mg\,(CO_3)_2$), chemical weathering dominates and is assisted further by physical erosion. Tropical countries either as islands formed from ancient coral reefs or which have been fringed with coral reefs will have much of their subsurface carbonate compressed to limestone rock. These ancient coral reefs are often covered with other sediment and then uplifted as part of the mountain-building process to well above sea level. Today it is possible to find an ancient coral reef with all its usual tropical features now as a limestone complex well inland and at a great height.

2.2 Karst Topography

There are many places in the world where there are large areas of limestone, dolomite and marble. Exposed to the surface or near the surface, these areas are chemically weathered and then eroded to form a distinct landscape called **karst** topography.

This name is derived from the German, *karst* for the Kras region, a limestone plateau surrounding the city of Trieste in the northern Adriatic and now located on the border between Slovenia and Italy. Karst areas are rugged, consisting of bare limestone cliffs, gorges, steep pinnacles and **dolines** (or sinkholes), often with large entrances to extensive underground cave systems.

Figure 2.1: Karst pinnacles, in Quangxi Province, south eastern China

Figure 2.2: Rillenkarren are the fluted grooves found on surface rocks in karst regions. They are formed as the slightly acid rainwater trickles down the rock. Wee Jasper Caves, New South Wales, Australia

Silica impurities such as mud are left behind within the large caverns and cave tunnels formed as the calcium carbonate the limestone rock is dissolved. Most of the limestone outcrops in the world were formed originally as coral reefs in warm, tropical seas. These reefs were made from the biological action of living coral polyps, which make their calcium carbonate exoskeletons from carbonate dissolved in the seawater. When the coral dies, more corals build up upon the old layers. Eventually these reefs become compacted into extensive limestone deposits. These can then be uplifted above sea level as limestone rock and eroded. As the limestone is formed by compaction, it cracks into a series of horizontal planes separated by near vertical joints – almost like a large brickwork pattern. Rain and river water will flow down these exposed vertical cracks and due to its slight acidity (of carbonic acid), will begin to dissolve the rock and enlarge the cracks.

Figure 2.3: Joints in the coastal limestone plateau of The Barran, County Clare, Ireland.

2.3 Limestone Caves

Limestone caves are formed if there is sufficient water available, by the dissolution of the rock along the bedding planes and joints. Most of their enlargement occurs below the **water table**, which is the top of underground water within the rock or soil.

Here the acidic water has a larger volume, greater surface area of contact and time to dissolve the rock. Similarly, caves can be formed in other carbonate rocks such as marble and dolomite.

As the carbonate rock is dissolves, it forms a solution of calcium hydrogen carbonate, which is then removed from the system as the water flows through the cave or evaporates:

$$CaCO_3 + CO_2 + H_2O \longrightarrow Ca(HCO_3)_2$$

calcium carbonate + carbon dioxide gas + water → soluble calcium hydrogen carbonate

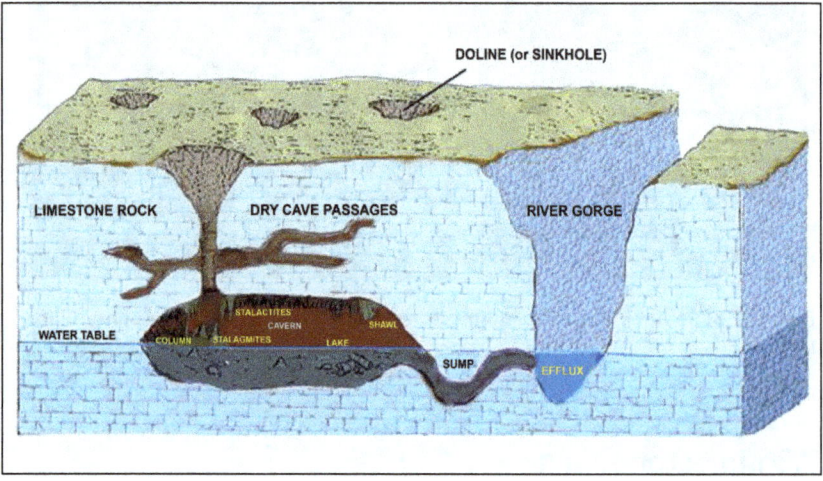

Figure 2.4: Diagram showing a typical pattern of carbonate rock weathering and erosion

Within the cave system, dripping and slowly flowing water, will evaporate if there is sufficient air circulation and new calcium carbonate, as calcite, is deposited as a **spelaeothem** or cave formation:

$$Ca(HCO_3)_2 \longrightarrow CaCO_3 + CO_2 + H_2O$$

soluble calcium hydrogen carbonate → calcite + carbon dioxide gas + water

The water evaporates and the excess carbon dioxide gas is liberated and then contributes to more formation of carbonic acid by dissolving in more water.

The most common spelaeothems are:

- **Stalactites** which hang from the ceiling as water drips down through cracks and deposits a thin circular rim of calcite crystal. This continues to grow in length to form straw stalactites, which look like crystal drinking straws. These eventually block so the mineral solution trickles down the outside of the straw, enlarging it into the familiar conical shape of the stalactite (memory hint: stalac<u>tite</u>s hang <u>tight</u> from the ceiling).

Figure 2.5: Stalactites at Jenolan Caves NSW, Australia

- **Stalagmites** form on the solid floor of the cave due to the dripping of water from above and they tend to be more rounded and wider in shape than stalactites. However, if the rate of water dripping is too fast, then there may not be a corresponding stalactite above. They may have a small indentation in their top or they may be pointed as the water has run down the sides.

Figure 2.6: Stalagmites at Jenolan Caves NSW, Australia. This formation is likened to a bishop on top of his pulpit (at left)

- **Helectites** (or cave mysteries), are stalactite-like material which grows out from a wall in any direction, forming thin, weirdly twisted tubes. How helectites are formed is controversial, but hypothesis is that the twisting may be due to an off-centred growth of crystals.

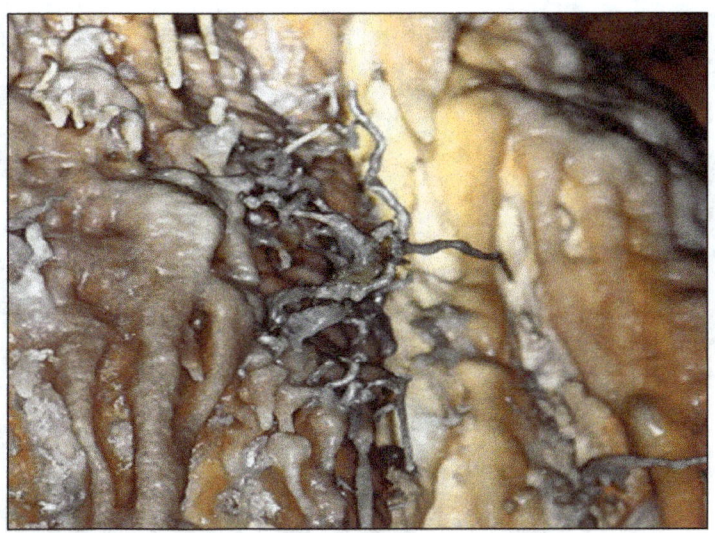

Figure 2.7: Helectites or Cave Mysteries at Jenolan Caves NSW, Australia

- **Rimstone pools,** or gours, are walled pools, often found on the floor of a cave below an active dripping zone. Starting from a simple pool of water, a calcite rim forms around the edge as the water evaporates. On sloping surfaces, these walls can build up to many tens of centimetres forming a semi-circular basin of some depth.

Figure 2.8: Small rimstone pools on the floor of a cave at Jenolan Caves NSW, Australia

- **Columns** form when an active stalactite joins with its corresponding stalagmite below.

Figure 2.9: Columns at Jenolan Caves NSW, Australia

- **Shawls** or curtains are very thin, often only a few millimetres thick and translucent. They are formed as ribbons of calcite crystallise as the mineral solution slowly trickles down a sloping wall.

Figure 2.10: Shawls at Jenolan Caves, NSW, Australia

- **Dripstone**, or flowstone, is a general term for calcite formations spread over a wider area where water has flowed over existing surfaces. Sometimes these may look like crystalline waterfalls.

Figure 2.11: Dripstone, Jenolan Caves, New South Wales Caves, Australia

 Online Video 2.1: Venture underground into a limestone cave
Go to https://youtu.be/OCbccRGRh84

Chapter 3: The Work of Moving Water

3.1 Introduction

Whilst there is some break down of rock by the sheer **hydraulic pressure** of large volumes of fast-moving water, most of the dramatic erosion by moving water is due to abrasion by the particles carried within the water. This causes large sections of rock and soil to be undercut, transported and then deposited elsewhere. Particles (**clasts**) carried by moving water are usually rounded because they are continually rolled around and their edges are smoothed off during transportation. The faster the movement, the bigger the particles which are transported, so in fast-water environments, such as in mountain streams or on storm beaches, clasts could be metres across, whereas in slow-water environments, such as in lagoons, lakes, slow-moving rivers and deep oceans, clasts may be very small.

The graph (below) is a modified version of the famous **Hjulström Diagram**, showing the velocities required to erode, transport and deposit clasts within a river. This graph however, only gives an approximation of real conditions, and does not take into account the depth of the water nor acceleration and de-acceleration effects.

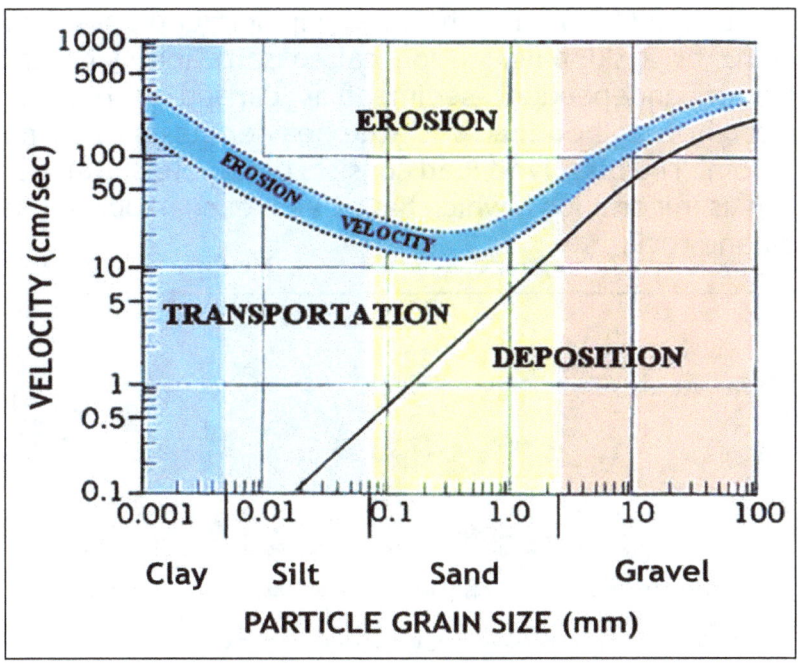

Figure 3.1: Particle size and stream velocity (After Åke Sundborg, *Geografiska annaler*, häfte 2-3/1956, pp. 125-316)

From the graph, for example, a good sized particle of about 10 millimetres in diameter, would require a stream velocity greater than 100 centimetres per second (about 3.6 km/hour or a little over 2 miles per hour) in order to be transported. Moving water has produced two major landscapes of erosion: river landscapes and coastal landscapes.

3.2 River (Fluvial) Erosion and Deposition

Particles within a stream move by different processes depending upon their size. The larger particles are moved along the bed of the stream in horizontal pushing, rolling or a skipping motion called **saltation**, whereas smaller, independent sediment is carried within the water flow as **bed load** with the heavier clasts near the bottom. The **dissolved load** consists of soluble materials, such as mineral ions, which have been leached out of the soil and rock.

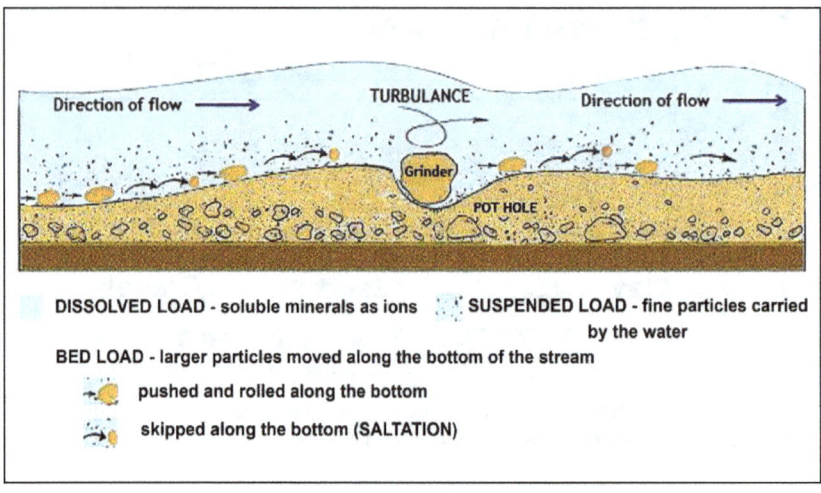

Figure 3.2: A diagram showing the loads carried by a stream

Water flowing downhill from highlands having high precipitation as rain or snow, will take the easiest course, often as a single, twisting, sinuous channel. Near the source area, the **headwater tract**, the flow will be greater, the pathway generally straight and the particles in the river larger as boulders.

Figure 3.3: The Rio Colca, western Peru near its headwaters. Notice that the river only bends slightly, and that the ridges (spurs) coming down to the stream barely overlap.

Figure 3.4: Large boulders in the canyon of the Rio Colca. The largest here is about two metres across and all are only slightly rounded.

Online Video 3.1: Take a dangerous trek down the 4000 metre deep Colca Canyon, western Peru. Go to https://youtu.be/PwRYBITmWZw

The shape of the river valley here is a typical V-shape with the **overlapping spurs** or descending ridges, relatively close together because the river does not bend or **meander** very much. The sides of the valley are very steep, sometimes becoming near vertical forming a gorge. Waterfalls and rapids are also often a feature of the headwater tract.

Figure 3.5: The Barron Falls in northern Queensland, Australia

Figure 3.6: Approaching the lower section of the Iguaçu Falls on the Brazil-Argentine border.

 Online Video 3.2: Visit the Iguaçu Falls on the Brazil-Argentine border. Travel by high speed boat into the base of the falls and visit the Devil's Throat. Go to https://www.youtube.com/watch?v=qycwdJz9BJc

Further downstream in the **middle tract**, the slope of the valley is gentler, the stream flow less rapid, the overlapping spurs and scree slopes broader. The clasts in the river bed become smaller and more rounded. In addition, as the valley gets broader and the river continues to cut down its floor, stepped **river terraces** may form above the river's banks on either side.

Figure 3.7: Large river terraces on the Rio Colca near the village of Maca, Peru. They have also been terraced in the Incan- style for farm irrigation

Figure 3.8: The Gross Valley in the Blue Mountains of Australia has been formed within a dissected plateau as this sandstone range has been rapidly uplifted as the river eroded the valley

In the middle tract, transported particles are deposited as river gravels in **point bars** on the inner edge of the meanders, and as **longitudinal bars** in the centre of the stream where a straight section is called a **reach**. Often longitudinal bars are slightly tear-dropped shaped along their axis, with a bulbous end trailing off to a long tail downstream. These are **lag deposits** formed by a sudden increase then decrease in the flow rate which will wash coarser gravels downstream during floods.

Figure 3.9: A reach of the Waimakariri River, South Island, New Zealand

As the river descends onto the coastal plain at its **coastal tract**, the meanders become very broad and twisting, often changing their direction and leaving bends cut off from the main stream. These are called **oxbow lakes** or billabongs. There are also many tributaries or secondary streams joining the main stream along this tract.

Sediments in the very slow-moving stream and on the banks are fine muds and silts. Sometimes these build up within the stream where water flow has decreased, perhaps at an obstruction, forming mud bars or flats. Lake environments are termed **lacustrine** and swamp environments are **paludal**.

Figure 3.10: Aerial view of the meanders of the Río Paraná Argentina

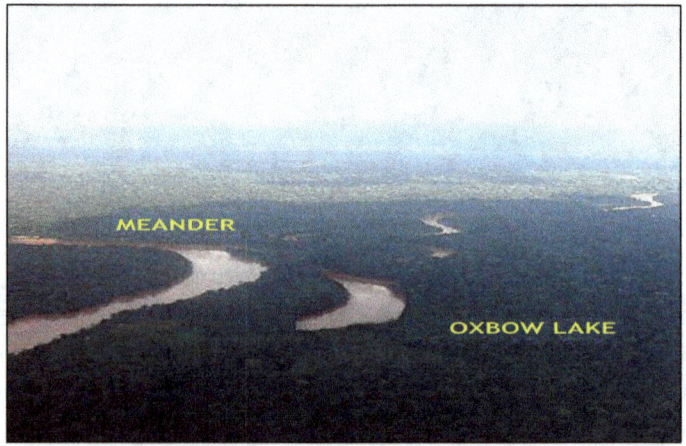

Figure 3.11: An oxbow lake (Lago Sandoval) near the Madre de Dios River, a tributary of the Amazon River, Peru

Figure 3.12: A white caiman sunning itself on the mud of the slow-moving Rio Madre de Dios, Peruvian Amazonia

 Online Video 3.3: Travel down the Madre de Dios River, a tributary of the Amazon River in a Peki-peki (motorized canoe) and then trek inland to Lago Sandoval.
Go to https://www.youtube.com/watch?v=VhJ7Ve1FbL0

At the end of the coastal tract, the river runs into the sea by several **distributaries** which are smaller branches formed as the river splits. Where the coastal plain is very flat, sediment may build up out into the sea and form a **delta** of many distributaries.

Figure 3.13: A delta region south of Brisbane, Australia. Notice the many mud banks and islands which have been formed here.

Here, the sediments of the river have progressed from sand to fine sand then muds and then very fine silt. Often the silt builds up and forms bars and mud banks, with smaller islands within the broad flow region of the river. The sides of the river along this tract build up as muds, but as the river continues to cut down its bed, the banks become raised. These raised banks are called **levees.** These slope away from the river to the **flood plain** beyond, which is usually lower than the river bed and so subject to regular flooding when the levees are breeched. The water of the river here is usually saltwater and tidal, and often the sediments and any fossilised life they contain, may show a mixture of freshwater and marine types.

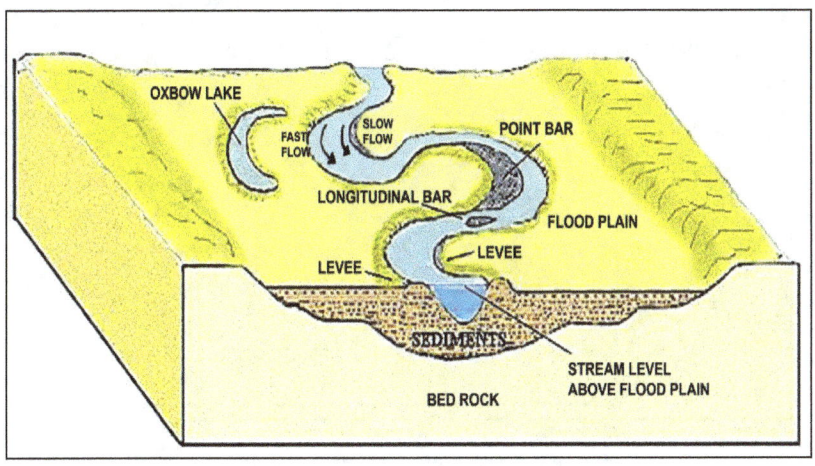

Figure 3.14: A meandering stream system

Where the stream gradient is uniformly steep, and there is plenty of sediment, a **braided stream system** may develop. In this type of river system, there are many channels which constantly change their direction, joining with other channels and making new temporary islands, lakes and bars. Where the water flow is faster, gravel bars dominate over the finer sand bars. These braided streams are often found at the foot of mountains within alluvial fans, as outwash streams at the end of glaciers, and within river deltas and other uniformly flat plains called **peneplains.** When the stream system contains less sediment and the slope is moderate, it may consist of only a few deep channels which may join and form an **anastomosing stream** system.

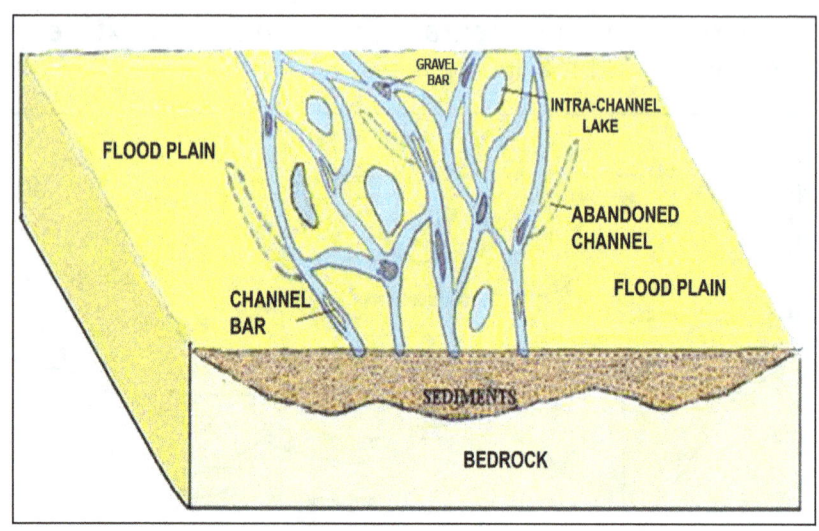

Figure 3.15: A braided stream system

Figure 3.16: An outwash stream from the Fox Glacier, New Zealand.

3.3 Coastal Erosion and Deposition

The interface between land and sea is a complex struggle between the erosion of the land and deposition of new material by the sea. The amount of erosion by the sea depends upon the force of the wave action, which depends upon the strength of the local prevailing winds, their direction, the steepness of the shoreline, and the hardness of the rocks of the shoreline. Erosion at the shoreline is caused by the:

- Hydraulic pressure of the moving water coming in from the sea and expanding cracks within the rock below the waterline. This undermines the coastal rock and causes debris to fall to the base the cliff thus formed.

- Wave action due to the many particles picked up by the water movement forward. This is a rolling motion which causes the particles to grind down the basement above the bottom of the waves (or wave base). If the basement is shelving, that is, getting shallower, the volume of water will be compressed into a smaller space, so that it rises up and forms cresting breakers. These too, will produce a dramatic effect on their region of impact.

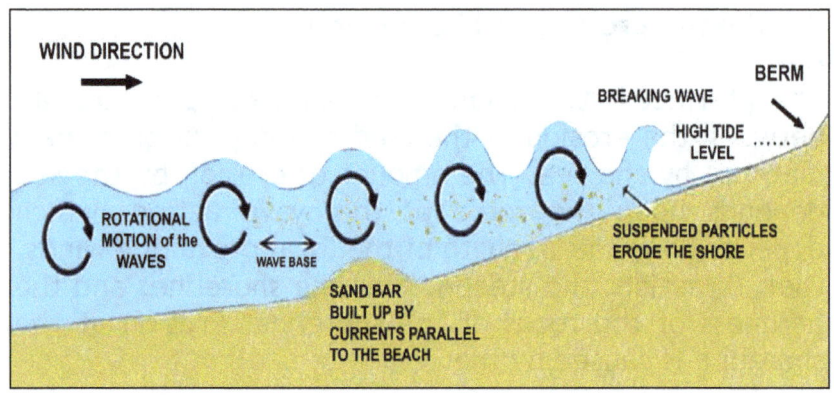

Figure 3.17: The effects of wave action on a shelving shoreline

If the prevailing winds and the waves they generate, strike the shoreline at an angle, the waves wrap around headlands and scour out shapes which look like large fish-hooks from above. When sand builds up on these curves they form **zeta-curved beaches**, so named from the shape of the Greek letter, Zeta. The sand is brought into these indentations by the **longshore current** which runs parallel to the coastline and is caused by the waves continually being forced around the headlands. Along the eastern Australian coast, for example, this current hugs the coastline and comes from the south because the prevailing winds are from the southeast as the southeast trade winds. In California, where the prevailing winds are the northwest trade winds, the longshore current flows south and the zeta-curved beaches also trail off in that direction.

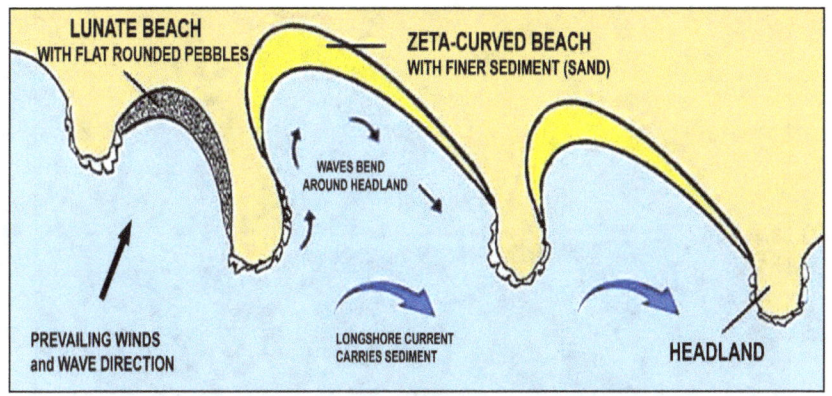

Figure 3.18: Formation of zeta-curved beaches

Where the waves strike the shoreline directly, that is at ninety degrees to the direction of the coastline, the wave action may be more severe and only the larger particles are able to be deposited and form a **storm beach.** These usually have a lunate or moon-like shape, and their clasts usually consist of larger pebbles which are flattened and rounded and often are aligned and overlap each other. Such an arrangement of clasts is called **imbrication** and is caused by the to-from action of strong waves lapping onto the shore. The pebbles overlap and align themselves – usually with their longer axis facing up the beach allowing less friction and greater stability.

Figure 3.19: A storm beach of pebbles, Half-Moon Island, Antarctica. Note the rough nature of the surface, including several changes in angle (berms) and the rounded and flattened pebbles (foreground). The Gentoo Penguins were most curious!

In warm, tropical regions where the sea conditions support coral as extensive reefs, beaches may consist of broken fragments of the inner reef washed up to form coral beaches.

Figure 3.20: A coral beach with detail (insert) of the broken nature of the beach material, Moso Island, Vanuatu, South Pacific.

Other, more abrupt landscape features are produced by the extensive erosion of a rocky shoreline. These can include:

- Sea cliffs of vertical, bare rock which usually have flat tops and scree slopes below the high tide level.

Figure 3.21: Cliffs of Moher, County Clare, Ireland. They are over 200 metres high and are of hard shales and sandstone

- Headlands when the sea has eroded a rocky shore leaving a narrow strip of the hardest, most resistant rock.

- Sea arches when the sea erodes a tunnel completely through a narrow headland where there is a natural weakness or crack in the rock.

- Sea caves when the headland or cliff is only partly hollowed out by wave action and hydrodynamic pressure, especially along a natural weakness in the rock face.

- Sea stacks when sea arches collapse or when the outer section of a headland has been totally cut off from the mainland. On a larger scale, this section of land may form a significant island just off the shore.

Figure 3.22: Limestone sea stacks of the Twelve Apostles, Southern Ocean, Victoria, Australia

- Marine rock platforms form if that part of the coast is slowly being uplifted by vertical earth movements as part of the maintenance of surface equilibrium or **isostasy**, and the eroded base of the headland is permanently lifted out of the water above the high tide level.

Figure 3.23: Headland and rock platform (centre right) of sandstone at Caloundra, Queensland, Australia

Figure 3.24: A tessellated pavement – a rock platform which has been cracked by many fractures, in-filled with a harder material which stands out as the softer rock between the joints erodes. Queensland, Australia

Figure 3.25: The Blowhole, Kiama, New South Wales, Australia. The sea has eroded a tunnel in the basalt rock platform and has undermined a jointed section above which now forms a hole through which the sea surges

Figure 3.26: Sandy beach with berm (right), Greenmount Beach, Queensland, Australia.

Other landforms are produced by the **deposition** of sediment, especially sand, at places where the longshore current or sediment flow is interrupted. Such landforms include:

- **Sand spits** or long arms of sands which often curve around the mouth of rivers and at some quieter headlands, with the deposition more pronounced where the sand-flow first comes into contact with the obstruction.

- **Tombolos** (sand-tied islands) are formed when the shallow section between a near-shore island and the mainland fills with sand carried in by a current. The sand spit formed between the island or sea stack, grows higher when the finer sands are uncovered at low tide and then blown higher by the wind up onto the spit. Many of these tombolos are isolated by water at high tide.

- **Barrier islands** may be up to 200 m in height and are formed by extensive wind deposition of sand offshore and parallel to the coast. They are built up when a strong longshore current encounters an underwater obstruction, such as a submerged rock outcrop. At low tide, the uncovered sand is blown up by the prevailing winds and gradually this accumulates into a large sand island. These may be many kilometres long and usually parallel to the shoreline, often forming a large, semi-enclosed bay e.g. Moreton Bay, Brisbane, Australia and Galveston Bay, Texas.

Figure 3.27: The long climb up the sand of Moreton Island – a barrier island of sand rising to over 200 m – off the coast at Brisbane, Australia

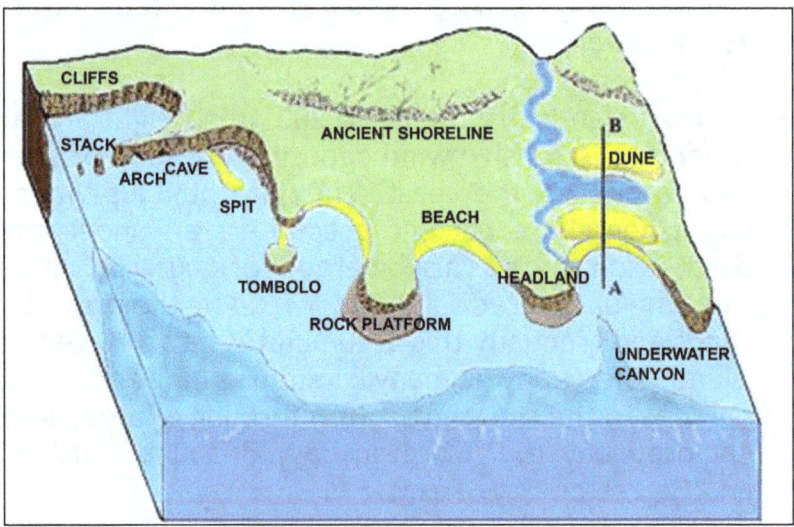

Figure 3.28: Some of the features of a coastline

A particular feature of many beaches is the **dual barrier system,** consisting of a series of parallel, long sand dunes just inland from the beach. At low tide, the finer beach sand is blown across the beach above the tidal cutting edge to form a frontal dune, or fore dune behind the beach. Finer material from the top of this dune, or even from the beach in strong winds, will continue to travel in a wind-blown arc further inland to form another line of dunes called the hind dunes. Between these parallel dunes, and sheltered from the wind, is a long hollow called a **swale**. This may fill with water to form a lake or swamp in which organic matter may accumulate from the dead plants which may grow there. In time, further growth and compression by more sediment on top may convert this organic matter into a sandy peat called **beach rock** or coffee rock. This can be exposed if the beach and its dune system are further eroded.

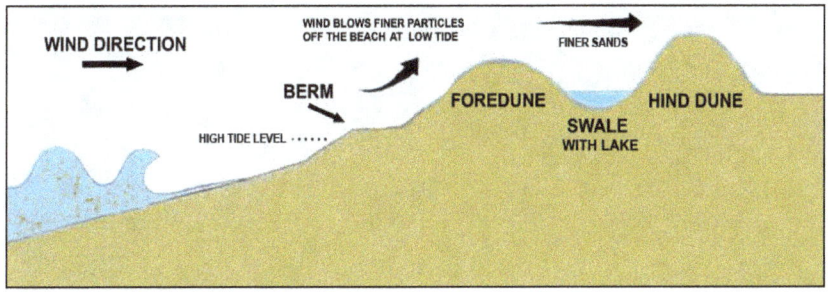

Figure 3.29: Diagram showing a typical dual barrier system of coastal dunes

Deposition of material will occur only in places along the shore where the wave action is minimal and currents can deposit the sediments formed elsewhere by erosion. Materials inshore may be coarse, depending upon the severity of the wave action, but offshore the sediments change from sand to fine sand and then muds and silts. Often the longshore current will carry sands just off the beach, but parallel to it and refract around into bays and river mouths forming sand bars.

 Online Video 3.4: Sand bars form as the longshore current carries sand around barriers and into river mouths and bays. https://youtu.be/GZWpUA9BWrw

Chapter 4: The Work of Moving Ice

4.1 Introduction

There are many parts of the world where moving ice produces most of the landforms. Other parts of the world with much flatter relief and little snow and ice may show traces of ancient glaciation in relic structures and glacial sediment. Glacial erosion is caused mainly by the abrasion of embedded rock material within the ice. These rock fragments come from the action of frost wedging in the cliffs above the glacier. Here, water seeps into cracks in the rock formed by expansion and contraction during day and night cycles. This water freezes and as ice crystals form, it expands and pushes the rock apart. It is this broken rock which falls down into the fresh snow which does much of the abrasion when the snow is compacted into ice.

Figure 4.1: Rock material (moraine) embedded in the Franz Josef Glacier, New Zealand

There are two main types of glaciation, alpine and continental.

4.2 Alpine Glaciation

Alpine glaciation is the more well-known of the two glacial environments, and occupies the high mountainous regions of the world. Whilst ice is softer than rock, freezing water can exert great pressure by expanding in cracks in the rock and breaking off large pieces of a cliff face – usually with a dramatic, thunderous noise. In alpine areas, where bare rock is the usual feature, frost wedging and expansion by temperature differences will constantly break off large amounts of irregularly-shape rock which may then fall into snow fields below. As this rock-filled snow is compacted by further snowfalls and low night-time temperatures, it turns into ice which slowly moves downhill under gravity. It is the large amount of rock debris **(till)** within the ice that erodes the country rock by abrasion in a scraping and sliding motion on the sides and the floor of the new U-shaped valley thus formed.

Figure 4.2: Looking up the Franz Josef Glacier, New Zealand

Figure 4.3: New piles of rock on the top of the Franz Josef Glacier, New Zealand. They will be embedded with the next snowfall

Figure 4.4: Large, irregular rocks (of schist) of the Franz Josef Glacier, New Zealand

Figure 4.5: Large striations or scratches on the wall of the valley of the Franz Josef Glacier caused by the grinding effect of the rock embedded in the passing ice

Usually at the head of a glacier is a large snowfield called a **névé**, which continuously supplies new ice with every snowfall. The surrounding cliffs also provide this snowfield with broken rock. As the glacier moves downhill, it rides up and down over the irregular curves of the rock floor below. Even though ice can be considered as having the ability to flow, called **rheidity**, it does act as a rigid structure when it is made to bend over the uneven valley floor. As the glacier bulges upward, it fractures in linear sections forming deep **crevasses**, which may go to the bottom of the glacier hundreds of metres below. When the glacier is bent downwards it also fractures, but the segments thus formed are squeezed upwards as huge wedges of ice called **seracs**. Melting water on the surface of the glacier often enlarges cracks or holes in the ice by their flow, forming circular swallow holes or **moulins** – a term from the French meaning windmill, as the water often races around the hole like water going down a kitchen sink.

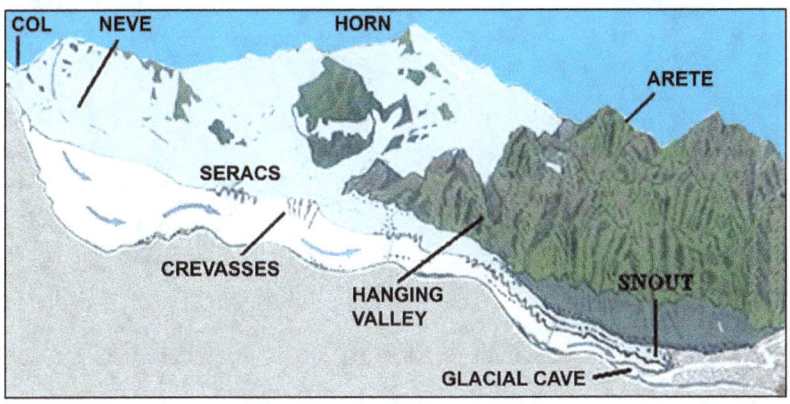

Figure 4.6: Diagram showing major features of a glacier

Figure 4.7: Crossing a small crevasse in the Franz Josef Glacier, New Zealand

Figure 4.8: heading into the snowfield of seracs, of the Franz Josef Glacier, New Zealand

Online Video 4.1: Trek across glaciers in Switzerland and New Zealand.
Go to https://youtu.be/zo1bQLLArlM

Glaciers advance and retreat depending upon the amount of snowfall upon the neve. If a glacier completely retreats, then many of its erosional features become apparent. These include:

- **U - shaped valleys** which are formed because the action of erosion here is broken rock within ice being pushed as by a grader downslope. The valley floor is flat and covered with broken rock material and often a braided stream system. The walls of the valley are often vertical and show smooth faces with extensive scratch marks called glacial **striations**. Where these valleys come down to the sea, a **fjord** will form as a long, narrow, vertically-sided inlet.

Figure 4.10: Glencoe Pass, Scotland is a typical glacial valley

Figure 4.11: Part of the Sognefjord, Norway. It is a deep, U-shaped valley, well inland but entered by the sea

- **Hanging valleys** occur when a tributary glacier meets the main glacier. When both retreat, the tributary glacier valley is above and hanging over the main valley. Often there will be a waterfall coming off this upper valley.

Figure 4.12: Waterfall from a small hanging valley – on the route of the Flam Railway, Norway

- **Moraine** is the angular rock which is piled up along the sides of the glacier as lateral moraine, at the bottom of the glacier as ground moraine, or at the end or snout of the glacier as terminal moraine. If two glaciers meet and combine, then the lateral moraines of each glacier will also combine at the middle of the new and larger glacier as medial moraine.

Figure 4.13: Lateral moraine (foreground) and medial moraine of the Gorner Glacier, Switzerland

- **Arêtes** are the sharp ridges which form as ice erodes both sides of the ridge.

- **Truncated spurs** are formed when a glacier erodes past these arêtes or ridges and cuts of the ends at the u-shaped valley.

- **Horns** or peaks are the sharp, pointed mountains of the most resistant rock.

Figure 4.9: The Matterhorn, a famous alpine peak on the Swiss-Italian border

- **Cirques** are the small, deep rocky hollows left at the head of the glacier once it has retreated. It was formed originally below the neve as the weight of the ice gouged out the rock before starting its flow downhill. Often they have a small, but deep lake in the lowest part.

Figure 4.14: A large cirque lake formed from a glacier which once existed here in the Andes south east of Santiago, Chile. The smaller hanging glacier at the rear of the lake is the Morado Glacier

- **Col** or pass is a low gap between peaks or ridges. In some places, they may be called a saddle if broad and relatively flat and between the mountain peaks.

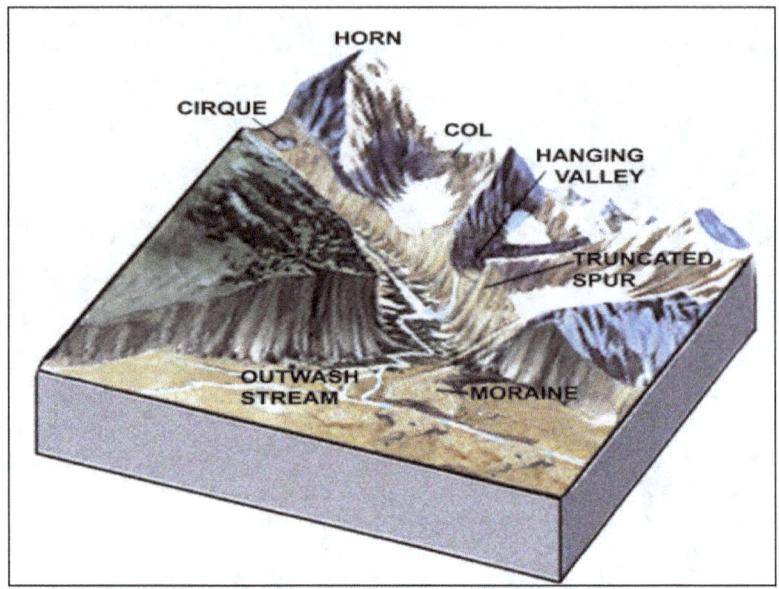

Figure 4.15: Diagram showing some of the features left by alpine glaciation

A glacial valley may broaden and extend out from the mountains onto a flat plain called a **piedmont plain,** from the Italian meaning: at the foot of the mountains. Other features may form due to the dumping of glacial moraine. These become apparent after the glacier has retreated. These features include:

- **Eskers** which are long, sinuous mounds of stratified or layered moraine and are thought to have been left by the meltwater stream within the ice cave below the glacier.

- **Drumlins** from the Irish word *droimnín* - littlest ridge, are small, elongated hills which are bulbous at one end and made of glacial till. Their tapered, elongated axis is in the direction of the ice flow and they may occur together as a drumlin swarm. They are also probably formed below the glacier.

- **Erratics** are very large, irregular boulders, often many metres across and precariously balanced upon smaller rocks. They have been carried in the ice some distance from their original source and so usually have little relationship with the rock upon which they have been dumped.

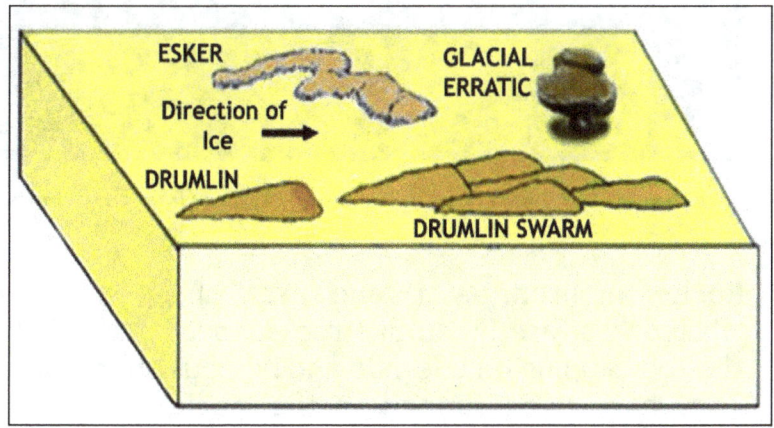

Figure 4.16: Eskers, erratics and drumlins - features left behind as a glacier passes

Figure 4.17: The Lakeland region of Finland has many lakes separated by eskers and drumlins

- **Rôche moutonnée** (French for sheep rock) or whalebacks, are hump-like rocks, usually rounded at the top, sloping on one side and bulbous on the other. They have been formed as the glacier grinds over bedrock, abrading the uphill end (the **stoss** side) and plucking broken rock off the downhill end (the **lee** side).

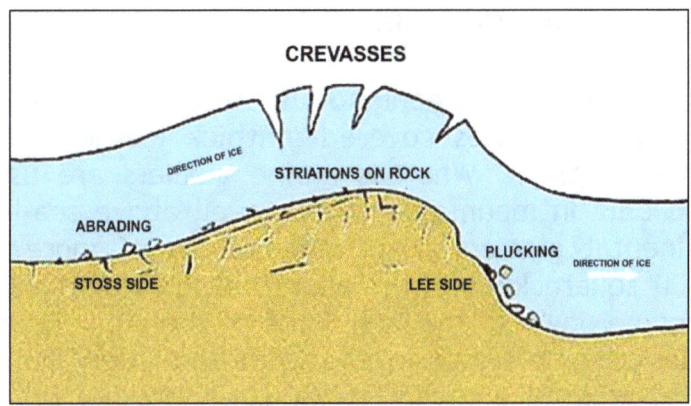

Figure 4.18: Diagram showing the formation of a rôche moutonnée

Figure 4.19: Glacial striations on bedrock, Cradle Mountain, Tasmania, Australia

4.3 Continental Glaciation

The continents of Greenland and Antarctica have most of their landmasses covered in thick icesheets many kilometres thick. Whereas alpine glaciers are usually found only in mountainous area, well above sea-level, continental glaciers cover huge areas of more than 50,000 square kilometres, near the poles of the Earth and come down to the sea. Some of the alpine glaciers of Alaska come to sea-level and those of New Zealand, the Franz Josef and the Fox Glaciers, are not far from the sea and have luxuriant forests of trees and ferns at their ends. Where large continental glaciers meet the sea, they often break off and form huge icebergs which may travel on ocean currents for many thousands of kilometres.

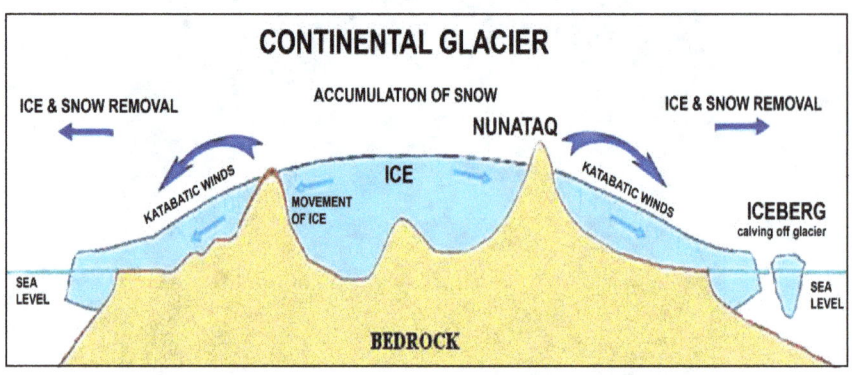

Figure 4.20: Diagram showing a continental region such as Antarctica

Antarctica is divided by the Transantarctic Mountains, one of the world's longest chains of mountains, which forms the boundary between East and West Antarctica. At about 3,500 kilometres long and varying from 100-300 km in width, they rise to over 4,500m above sea level.

Most of the mountain chain is buried beneath the ice with the peaks showing through as isolated **nunataks**. West Antarctica contains several large ice shelves, the Ronne-Filchner and the Ross Ice Shelves, which float out over the sea as the ice flows from the Transantarctic Mountains.

The Antarctic Peninsula is a long arm of mountains and glaciers which reach down to the several straits which separate the mainland from the parallel chain of islands which stretch to the north towards the Drake Passage and South America.

Figure 4.21: A map of Antarctica

East Antarctica takes up about two thirds of the area of the continent and consists mainly of a thick sheet of ice resting upon a compressed landmass over 2000 metres below. This part of Antarctica is also the driest, windiest (average over 35 km/hr) and coldest (average of -28°C in summer and – 60°C in winter) part of the continent. Antarctica can be considered a cold desert because very little moisture comes into the region from outside. At times, strong, cold **katabatic winds** from the Greek for *flowing downhill* blow from the interior down to the coast.

There are also several valleys free from ice, such as the McMurdo Dry Valleys. They are ice-free because they have been uplifted out of the ice sheet and have exceptionally dry climates.

Figure 4.22: A large glacier coming down to the sea at Paradise Bay, West Antarctica. Notice the rough front of the ice wall which breaks off, calving icebergs.

Figure 4.23: A medium-sized iceberg, Bransfield Strait, West Antarctica. One of the ship's rubber dinghies is at the far right.

Figure 4.24: A Leopard Seal on a small ice flow watches the sea begin to freeze, Wilhelmina Bay, Antarctica. The sea freezes at -2°C with thin, greasy patches of ice being a warning sign of major ice formation

Figure 4.25: Exploring Wilhelmina Bay, in a 5 metre Zodiac rubber dinghy.

Figure 4.26: Mid-summer at Paradise Bay, Antarctica with the Almirante Brown Scientific Station (Argentina) in the foreground.

Figure 4.27: The manned British base at Port Lockroy, Antarctica. A 1940's base now being restored by volunteers for tourist visits

Figure 4.28: Early radio room at Port Lockroy

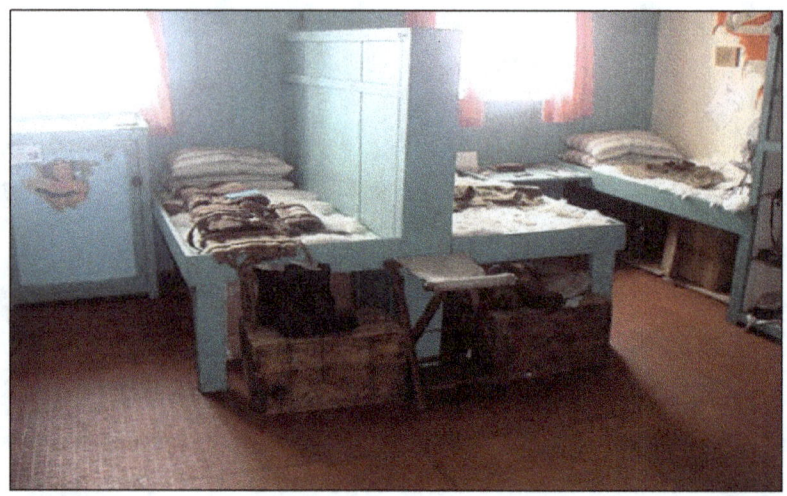

Figure 4.29: Early accommodation at Port Lockroy - wooden bunks with sheepskin covers. Heating is provided by a small coal-fire stove.

 Online Video 4.2 Explore the bleak Wilhelmina Bay, Antarctica and have a close encounter with Minki Whales.
Go to https://www.youtube.com/watch?v=Nt109CEDGnM

Chapter 5: The Work of Moving Wind

5.1 Introduction

Most of the common features formed by moving wind are found in the desert regions of the world. These are usually found between 30⁰ north and south latitude, although the deserts of the western Andes extend further south. Between these latitudes, there are high pressure zones which cause the winds to blow outwards, and

Figure 5.1: The beginning of the sandstorm season in Cairo, Egypt

so do not bring any moisture in from the sea. Some areas receive only hot, dry air from outside, such as the Sahara and Kalahari Deserts of Africa and the Great Australian Desert. Other desert areas are in the rain shadow of mountain ranges which trap any moisture coming from the sea such as inland deserts of the USA. Yet others are on coasts, where cold sea currents remove any moisture from the air such as the Atacama Desert of northern Chile and the Namib Desert of south western Africa. Some deserts, such as those in central Asia, including the

Gobi Desert, are simply too far away from the coast to receive any moisture.

5.2 Major Features formed by Wind

Landscapes of wind erosion, formed by moving wind are due to the nature of the rock surface in desert regions and the three processes of:

- **Deflation** which is the removal of loose material from the surface with smaller particles such as dust and sand requiring only modest winds.

- **Abrasion** is the eroding of rock and soil by the wind-borne particles which usually hug the ground depending upon the velocity of the wind.

- **Deposition** is the dropping of the material carried by the wind as these particles strike obstructions or as the wind decreases in velocity.

Deflation is the mass removal of fine silt called loess, sand and larger particles by the wind. It can produce large areas of scratched, bare rock called **desert pavements**, barren, flat rock surfaces with a layer of angular pebbles, including larger, triangular shaped boulders called **dreikanters** from the German for *three edged*. Rocks shaped by the action of abrasion by the wind are called **ventifacts**. Large and regular sandstorms are also a feature of desert regions.

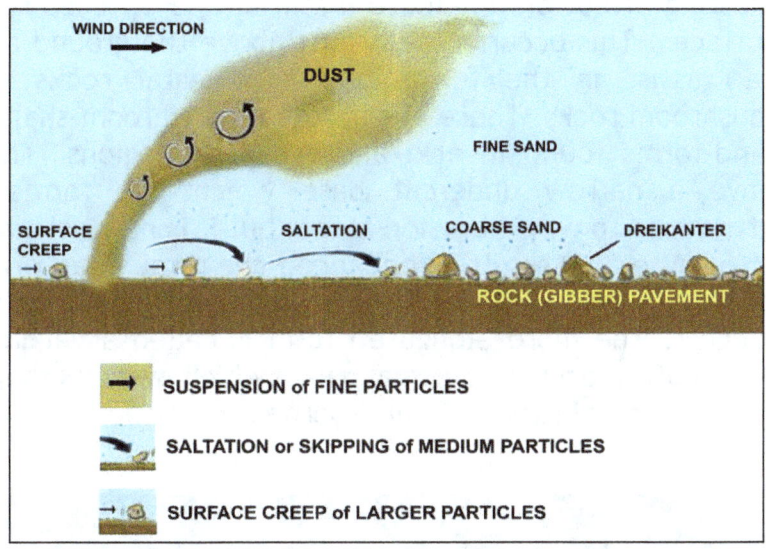

Figure 5.2: Diagram showing how the wind moves particles in the desert

Figure 5.3: Desert Pavement with some dreikanters, near Uyani, western Bolivia

Abrasion of the wind-borne particles causes a sand-blasting effect on any loose material and exposed rock surfaces. This occurs usually just above the ground and can assist in the formation of **pedestal rocks,** or mushroom rocks. These are unstable, mushroom-shaped land-forms found in arid and semi-arid regions. They have a narrow undercut base which was formerly attributed to wind abrasion alone, but it is now believed to also be as a result of enhanced chemical weathering at their base, where moisture would be retained the longest. The more elongated form is called a **yardang** which often occurs as a small ridge which owes its shape more to the abrasion of wind-borne sand.

Figure 5.4: Wind eroded pedestal rock, the *Arbol de Peidra* or Stone Tree, near Uvani. western Bolivia (Photo: Matthew Scott)

Deposition produces the more familiar desert features of sand dunes. Dunes are not as common as one might imagine, as they occur only where winds are consistent and from a prevailing direction. Often, they may form as the wind-blown particles meet an obstruction, or a place where the overall conditions allow for a drop in the wind velocity. Sandstorms are also a feature of desert areas. If the wind direction is constant, it will blow the lighter particles up and over the obstruction forming the typical dune with a long uniform slope on the windward or **stoss** side and a sudden drop onto a steep slope on the **lee** side where it is protected from the wind.

Figure 5.5: Wind eroded outcrop and yardangs (distant at right), near Uyani, western Bolivia (Photo: Matthew Scott)

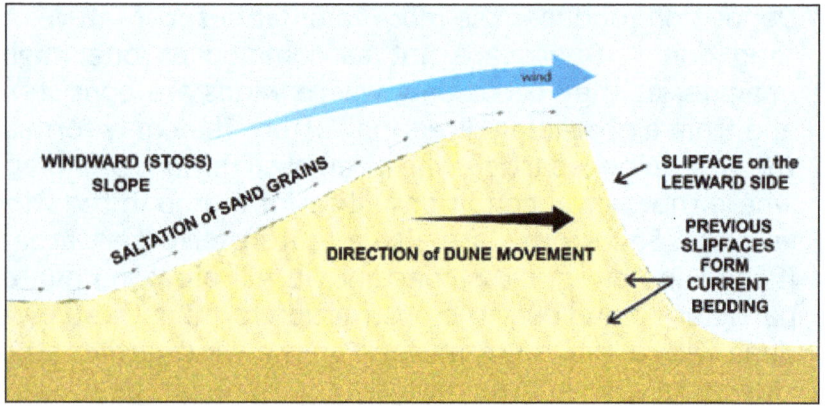

Figure 5.6: Formation and features of a single sand dune

Sand dunes may come in different forms depending upon the supply of sand and the nature of the prevailing winds. The main types of dunes are:

- **Barchan dunes** are crescent-shaped dunes and are formed in regions of limited sand with the wind blowing in one direction. The name barchan, is a modified version of the name *barkhan*, which was introduced in 1881 by the naturalist Alexander von Middendorf (Russian: 1815-1894) for the crescent-shaped sand dunes in Turkestan, and is Russian for dune.

Figure 5.7: A small barchan dune, Nubian Desert, Egypt

- **Longitudinal dunes** form where there is a moderate supply of sand and constant prevailing winds. These dunes are usually parallel to the direction of the wind.

- **Parabolic dunes** form as vegetation stabilises one end (or horn) of the dune, allowing the rest of the dune to migrate further. These are often seen in the dunes of coastal systems.

- **Transverse dunes** are formed where there is abundant sand and the wind blows consistently from the same direction at ninety degrees to the direction of the dune.

- **Star dunes** are pyramidal piles of sand with sharp tops which are formed when the wind blows from several directions, shaping the sand vertically rather than horizontally.

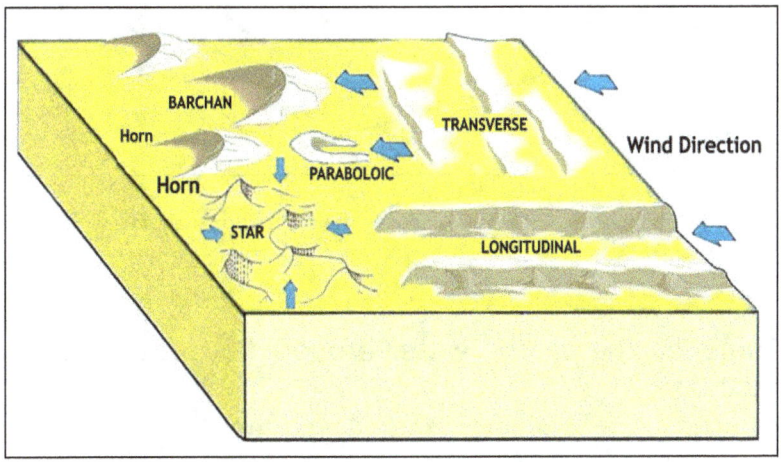

Figure 5.8: Diagram showing different desert dune systems

5.3 Major Features Formed by Wind and Water

Whilst wind erosion and deposition are the most consistent factors in desert landscapes, some other features are formed by very sudden and rapid flows of water. When it does rain in the desert, there is very little vegetation to hold sand, soil and rock material, and so there is massive erosion.

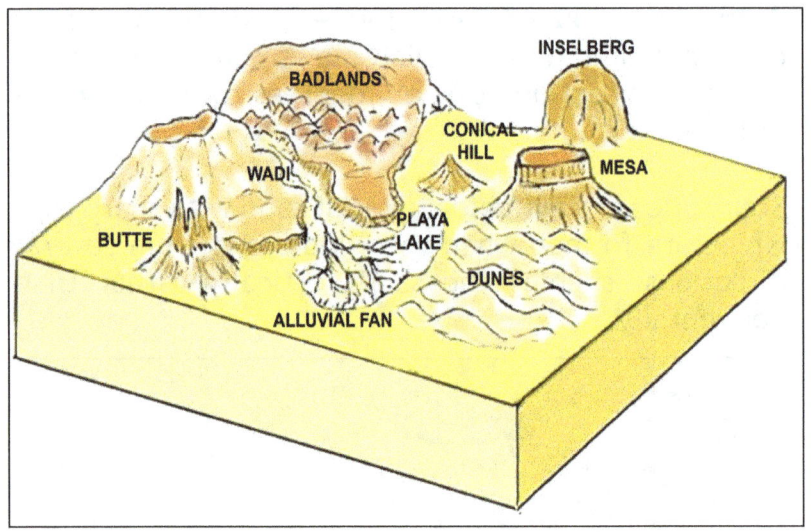

Figure 5.9: Diagram showing the main desert landforms

Some of the major water-eroded features of deserts include:

- **Wadis** or dry water courses which may be small gullies or very large valleys and broad depressions.

Figure 510: Aerial view of large, dry watercourses or Wadis, in the Sinai Desert, Egypt

Figure 5.11: A small wadi, Atacama Desert, Chile (Photo: Matthew Scott)

- **Mesas** which are small plateaux, having a distinctive flat top and steep sides which erode away to form **buttes** and conical hills.

Figure 5.12: A mesa in the Nubian Desert, Egypt

Figure 5.13: A conical hill in the Nubian Desert, Egypt

- **Inselbergs** from the German for *island mountain*, are large, isolated and rounded rocky mountains which usually stand alone on a desert plain. For example, Australia's Uluru, once called Ayer's Rock, is only a small fragment of an old pebbly sandstone plateau which has all but eroded to the surrounding flat plain.

Figure 5.14: Uluru, an inselberg in central Australia

- **Badlands** consist of badly dissected edges of hills or plateaux which contain many water-eroded spires and gullies.

Figure 5.15: Badlands with alluvial fans, Cafayate region, Argentina (Photo: Matthew Scott)

Some depositional features often associated with water include:

- **Alluvial fans** which form at the end of wadis or hills where small braided streams carry loose debris out onto desert plains.

- **Playas** from the Spanish for *beach* develop as temporary lakes when water runs off the surrounding hills into large, closed desert depressions. The water does not last long and evaporates quickly, often leaving polygonal mud cracks or mineral salt deposits. In some deserts, such the high desert of Bolivia, huge salt flats, called **salars**, such as the Salar de Uyuni cover over 10,000 square kilometres and are valuable source of salt (sodium chloride) and other minerals such as those of the element lithium.

Figure 5.16: Polygonal mud cracks in the playa at the Salar de Uyuni, Bolivia (Photo: Matthew Scott)

Figure 5.17: Harvesting minerals salts rich in the element lithium at an altitude of 3654 metres, Salar de Uyuni, Bolivia. (Photo: Matthew Scott)

Summary

1. Weathering is the chemical, biological and physical attack of Earth material - minerals rock and soil - where they are exposed, whereas erosion by water, ice, wind and gravity involves the mass movement of the material which has been broken down.

2. Physical weathering includes processes involving expansion and contraction due to temperature differences, including the cracking of rock by ice as it expands when crystallising from water.

3. Chemical weathering involves the change of minerals by naturally occurring gases such as oxygen, and oxides of carbon, sulfur, and nitrogen, and solutions of these gases which are acidic. Other acids are formed by the breakdown of organic matter in the soil.

4. Soils result from the weathering of rock, the accumulation of decaying vegetation called humus, and the chemical processes and leaching which can occur within the layers, or horizons of the soil profile.

5. Limestone regions (karst) are subject to considerable chemical weathering by the action of carbon dioxide and water (= carbonic acid), which completely dissolve the calcium carbonate of the limestone to form huge cave systems. Evaporation of the water content of the carbonate solutions this formed produces spelaeothems, or cave formations of calcite crystal, such as stalactites, stalagmites, helectites, columns, shawls and rimstone pools.

6. Weathering is more rapid in places where there are considerable temperature differences, such as in deserts both hot and cold, or where there is high temperature and high humidity such as in tropical regions.

7. Erosion, involving movement is more rapid if the surface has been badly cracked and broken by weathering. Erosion by water and ice is faster when the angle of the land is steeper, or the velocity of the wind or movement of the ice correspondently faster.

8. Moving water will erode rock material by hydrodynamic pressure but mostly by the rolling action of the particles within it. Water eroded particles tend to be rounded by such action, with the particles becoming smaller in size, more rounded and approaching spherical shape with greater transportation.

9. River erosion produces V-shaped valleys which have overlapping spurs and wide meanders as the river finds the easier pathway downslope. Young river valleys are relatively steep, often with gorges with few and smaller meanders. Mature river valleys are often broad, with wide, twisting meander loops, oxbow lakes and river terraces. Rivers break up into many distributaries if they travel across broad, sloping plains and form braided streams, or as they join the sea as deltas on broad coastal plains.

10. Coastal erosion is also the result of hydrodynamic pressure and the movement of eroded particles, but here the driving force is not the slope of the land but the strength of wind-blown waves and coastal tides. The angle of the waves directed onto the coastline, and their strength will determine the shape of the coastline, including cliffs, beaches and the size of rounded sands and gravels deposited.

11. Ice erodes because of the broken material (till) which is embedded the ice. Glaciers carve out large steep-sided U-shaped valleys which may continue to the coast to become fjords. Till may be piled up as moraine in front as terminal moraine, on the sides as lateral moraine or in the middle where two glaciers meet as medial moraine. Till is typically angular and varies in size from glacial flour to large boulders.

12. Deserts are subject to dramatic erosion because there is little vegetation to hold the soil. The temperature differences between day and night ensure a high degree of broken material through physical weathering. This can be used in erosion by abrasion of bare rock or deposited by wind as sand dunes. Water deposited sediment and minerals which has crystallised from solution, can form large alluvial fans and playas within land-locked deserts. The mineral deposits of these regions are a valuable source of salt and other chemicals.

Practical Tips

1. It is uncommon to find good, fresh rock surfaces which show minerals because rocks quickly weather. Keep a good lookout when collecting fresh rock specimens in new road or railway cuttings. Collecting in working quarries and private property will require permission of entry.

2. Karst regions are noted for difficulty in travel because of their badly dissected surfaces and concealed sinkholes. Cave exploration is a specialised activity, and it is a good idea to enlist the aid of guides or a caving club which will have specialised equipment, local knowledge and caving skills. Usually, a party will consist of no less than three persons with backup help if needed. Individuals should have good clothing including cold water wet suits, helmets, boots and at least three independent sources of light.

3. Erosion implies movement, so extra care is needed when walking on loose, steep surfaces. Talus slopes are dangerous and river gorges are often prone to sudden flash flooding, especially near the headwaters where steep sides limit escape. Always keep to higher ground, especially when setting up camp as rainfall is often unpredictable and will cause flooding well downstream.

4. Wave-cut rock platforms are a good source of geological data but wave action can also be unpredictable with the occasional rogue wave being a danger. Always keep a watch to seaward.

5. Glacial exploration needs specialised equipment such as crampons, ice poles, warm and waterproof clothing and strong boots, as well as local knowledge. It is always good practice to have a guide who can show the best route across the glacier and where to camp. Remember, glaciers are also moving!

6. Deserts require thorough planning for field trips. Hot and dry during the day, they are very cold at night. Even for a day trip, plenty (excess) of water, shady hat and loose clothing are needed. For overnight trips, one must consider the cold and exposure to the wind. Always log in and log out to a responsible authority. Judging distance is also a problem in the clear desert air, especially at high altitudes. The interesting rock formation which looks only a few kilometres away might be many times that distance and require more effort and time than is available.

7. In the field, a quick test for clay content in a soil is to wet (spittle will do) a sample of the soil on the palm of the hand and try to roll it into a sausage shape. Good clay content will allow this shape to be held up at one end without it breaking apart.

8. Black soil plains or other regions noted for high clay content are difficult to traverse during rainy periods. These areas are identified by their black, grey and red soil colours and telegraph and other posts at strange angles. In dry seasons, tall grass may hide large cracks which break car axels and the area should be avoided if rain is possible due to the potential to bog vehicles.

Multichoice Questions

1. A mineral commonly formed by <u>chemical</u> weathering is:

    A. Quartz
    B. Clay
    C. Orthoclase
    D. Hornblende

2. The table below shows a generalized relationship between temperature and humidity for various types of weathering:

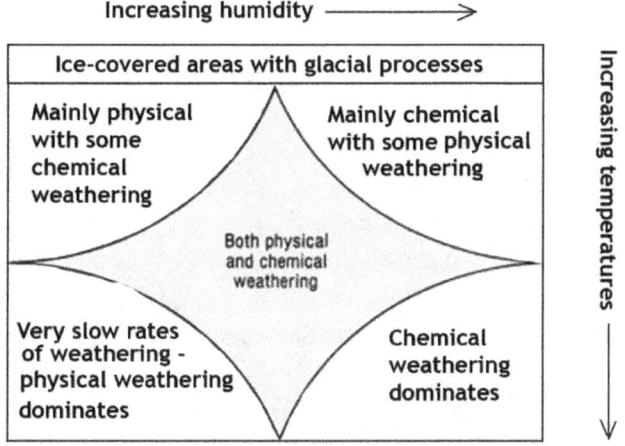

The graph suggests that:

A. The highest weathering rates occur in deserts
B. Chemical weathering is dominant in hot wet climates
C. Cold climates are mainly areas of physical weathering
D. Glacial processes are always accompanied by low humidity

3. Loose soil and rock material are collectively known as:
   A. Dirt
   B. Lithosphere
   C. Detritus
   D. Regolith

4. The action which is not an agent of erosion is:

   A. Running water
   B. Moving ice
   C. Chemical change
   D. Gravitational pull

5. The diagram below represents a view seen from above a river.

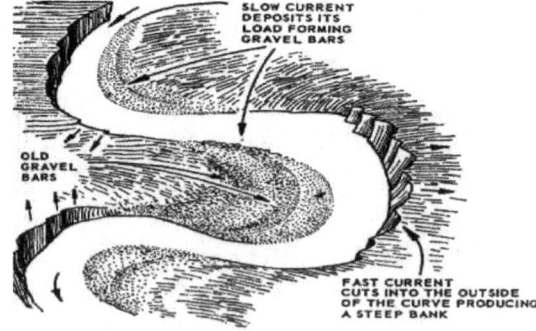

The whole structure featured is called a:

    A. Billabong
    B. Meander
    C. Levee
    D. Ox-bow

6. Karst is a term applied mainly to regions of:

    A. Limestone
    B. Basalt flows
    C. Sandstone plateaux
    D. Granite boulders

7. Till is a term applied mainly to regions of:

    A. Granite boulders
    B. Basalt flows
    C. Wind-blown dunes
    D. Glacial erosion

8. Sand grain sizes can be classified according to the following table:

$2mm - 1mm$	Very coarse sand
$1mm - \frac{1}{2}mm$	Coarse sand
$\frac{1}{2}mm - \frac{1}{4}mm$	Medium sand
$\frac{1}{4}mm - \frac{1}{8}mm$	Fine sand
$\frac{1}{8}mm - \frac{1}{16}mm$	Very fine sand

A sample of sand was passed through a sieve of hole-diameter size **0.25 mm.**

What would be the description of the sand sample that passed through the sieve?

        A. Fine to very fine sand
        B. Coarse to very coarse sand
        C. Medium to very coarse sand
        D. Medium sand

9. The following graph shows the predicted behaviour of sand particles of various sizes in a flowing stream:

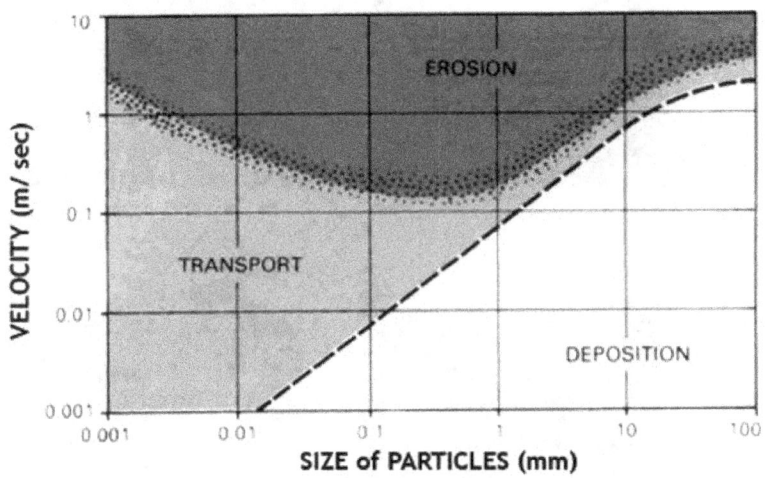

The approximate **minimum** velocity required to pick up particles of **coarse sand** (2mm diameter) would be:

    A. 0.001 m/sec
    B. 0.01 m/sec
    C. 0.1 m/sec
    D. 1.0 m/sec

10. The following photographs show various landscapes of erosion. Which one represents erosion by ancient glacial action?

A. A
B. B
C. C
D. D

Review and Discussion Questions

1. List the main effects that weathering and erosion has on humankind and habitation.

2. What might the typical soil profiles be like above the following bedrocks:

   (a) limestone
   (b) sandstone
   (c) granite
   (d) basalt

3. Discuss why there are a great variety of soil classifications.

4. Discuss the major similarities and differences in regard to processes, landforms and sediments of:

   (a) weathering and erosion
   (b) river erosion and coastal erosion
   (c) water erosion and ice erosion
   (d) water erosion and wind erosion

5. Discuss the importance of climate on weathering and erosion.

6. What are some methods used to prevent soil erosion in mountainous areas?

7. With the aid of a diagram, explain the nature and formation of:

(a) scree slope
(b) levees
(c) medial moraine
(d) tombolo
(e) zeta-curved beach
(f) yardang

8. Discuss the importance of water in the processes of weathering.

9. List some of the weathered products, such as minerals, solutions etc. of each of the following rocks:

(a) granite
(b) basalt
(c) limestone
(d) quartz sandstone

10. Rearrange the following minerals in their order of resistance to weathering, beginning with the one that weathers most easily. Account for each type of weathering:

quartz   anorthite   biotite   hornblende

11. Discuss the preparation and equipment needed to do research into the flow of water deep inside a cave within a karst area several hundred kilometres from your home.

12. What is a spelaeothem? Explain the factors which may cause the different types of these structures.

13. In a particular karst area, there is a suggestion that the limestone could be a valuable raw material. Discuss <u>for</u> and <u>against</u> the mining of this limestone.

14. What are the major hazards associated with exploring a glacier?

15. Playa lakes are most commonly found in desert areas and are important sources of minerals.
(a) how are these minerals formed? And
(b) give some of the uses of some of these minerals.

Answers to Multichoice Questions

Q1. B Q2. B Q3. D Q4. C Q5. B Q6. A Q7. D Q8. A
Q9. C Q10. A

Reading List

Allaby, M. (2000). *Basics of Environmental Science*. London: Routledge. 340 pp. ISBN-10: 041521176X

Anderson, R.S. & Anderson, S.P. (2010). *Geomorphology: The Mechanics and Chemistry of Landscapes*. Cambridge University Press. 651 pp. ISBN-10: 0521519780

Charlton, R. (2013). *Fundamentals of Fluvial Geomorphology*. London: Routledge. 320 pp. ISBN-10: 0415505518

Chesworth, Ward, (edit). (2008). *Encyclopaedia of Soil Science*. Dordrecht, Netherlands: Springer. ISBN: 1-4020-3994-8

Conway, A., R. Reynolds, R., Dise, N., Dubbin, B. & M. Gagan, M. (2006). *Air and Earth*. Milton Keynes, UK: Open University. 326 pp. ISBN-10 074926988X

Guinness, P. & Nagle, G. (2009). *IGCSE Geography*. London: Hodder Education. 216 pp. ISBN-10: 0340975016

Holden, J. (2011). *Physical geography: The Basics*. London: Routledge. 176 pp. ISBN-10: 0415559308.

Huggett, J. (2007). *Fundamentals of geomorphology*. London: Routledge. 480 pp. ISBN-10: 0415390842.

Migon, P (Edit.). (2010). *Geomorphological Landscapes of the World*. New York: Springer. 375 pp. ISBN-10: 9048130549

Walthan, T. (1974). *Caves*. London: MacMillan. 238 pp. SBN 333 17414 3

Willett, S. (Edit.). (2006). *Tectonic, Climate, And Landscape Evolution (Geological Society of America Special Papers)*. Geological Society of America. 434 pp. ISBN-10: 0813723981

Yerima, B.P. K.& Van Ranst, E. (2005). *Introduction to Soil Science: Soils of the Tropics*. Bloomington, Indiana. 440 pp. ISBN-10: 1412058538

Key Terms Index
(Page numbers in brackets)

abrasion (14, 101) is the grinding effects of rock fragments carried by water, wind and ice.

alluvial fans (111) form at the end of wadis or hills where small braided streams carry loose debris out onto desert plains.

alluviation (36) is the accumulation of regolith by deposition by water, wind or ice. In the case of soil profiles, it is the zone of regolith at the base of soil containing rocks and smaller particles accumulated on top of the bedrock.

angle of repose (5) is the minimum angle at which a slope will retain loose material.

anastomosing streams (64) form on more modest slopes and have several, well-defined and more constant, deeper channels which may be linked.

arêtes (86) are the sharp ridges coming down from alpine peaks.

avalanches (10) are formed when snowfields on the upper steep slopes of hillsides become unstable due to increased weight and collapse suddenly downhill.

badlands (110) consist of badly dissected hills or plateaux which contain many water-eroded spires and gullies.

barchan dunes (105) are crescent-shaped sand dunes mostly found in desert areas and formed from winds blowing consistently from one direction.

barrier islands (74) are formed by extensive wind deposition of sand offshore but in a shallow coastal region, often parallel to the coastline.

beach rock (76) or coffee rock is the red-brown organic, partly compacted sand formed within a swale between coastal dunes but often exposed by erosion.

bed load (55) heavier particles carried along the bottom of rivers.

Berm (figures pp.69 & 73) is the change of angle on a beach and represents the highest line of wave action. They have a steeper slope down to and below the tide level but level off above this where the beach is exposed to the wind.

biological activity (15) in physical weathering is the breakdown of rock by the action of living things e.g. tree roots and lichen.

Bowen's Reaction Series (for weathering) **(25)** shows the sequence of weathering for minerals which is a reverse of Bowen's Series for crystallization of minerals from molten rock.

braided streams (64) are found on relatively steep slopes, and form as several smaller streams which regularly change course, connecting and disconnecting with each other, depositing sediment across the width of their course.
buttes (109) are tall sharp or rounded spires found in desert regions due to excessive erosion of mesas.
carbonation (29) is the chemical combination of carbon dioxide gas and water to form carbonic acid which can then attack and dissolve carbonate minerals such as calcite and dolomite in rocks such as limestone.
carbon cycle (19) is the large recycling of the earth's carbon compounds, especially carbon dioxide in the upper crust, the atmosphere and the hydrosphere.
cations (29) ions or charged groups having an overall positive charge.
cirques (87) are the deep, circular lakes found at the exposed head of a glacier now removed.
clasts (53) are the particles in sedimentary rocks such as san grains in sandstone.
coastal tract (60) is the part of a river system near its end on the plain near the sea.
col (88) or mountain pass is a low gap between alpine peaks or ridges.
columns (51) are spelaeothems formed when a stalactite from the ceiling joins with a stalagmite from below.
cornices (10) are unstable, overhanging masses of snow on the upper slopes of alpine regions.
crevasses (82) are the deep cracks in a glacier formed as the glacier moves over an upward bulge in the bedrock below.
deflation (101) is the removal of loose particles by the wind.
delta (62) is the exit of a river into the sea by many branching distributaries separated by many flat islands.
deposition (74, 101) is the dropping of sediment because of the reduction in speed of the transporting medium such as water or wind, or the melting or retreating of ice.
desert pavements (101) are large areas of barren, flat, scratched, bare rock surfaces eroded by wind-borne particles and often with a layer of angular pebbles or dreikanters.
dissolved load (55) consists of the soluble materials carried within a river.

distributaries (62) are the branches which break off from a river on the flat coastal tract.
dolines (44) or sinkholes are the large, basin-like openings on the surface of karst regions due to the collapse of a cave roof below.
dreikanters (101) are large, triangular shaped boulders formed in deserts by sand abrasion and are named from the German for three edged.
dripstone (52) is a general term or any unnamed cave formation.
drumlines (89) named from the Irish word *droimnín* - littlest ridge, are small, elongated hills which are bulbous at one end and made of glacial till piled up below the moving glacier.
dual barrier system (76) consists of a series of parallel, long sand dunes just inland from the beach and having a depression or swale, between them.
eluviation (36) is the downward or sideways movement of water and suspended material down a slope or through a soil due to rainfall.
erratics (90) are very large, irregular boulders precariously balanced upon smaller rocks. They have been carried in the ice some distance from their original source and so usually have little relationship with the rock upon which they have been dumped.
erosion (2) is the abrasion and removal of rock and soil by water, ice, wind and gravity.
eskers (89) are long, sinuous mounds of stratified or layered moraine and are thought to have been left by the meltwater stream within the ice cave below the glacier.
exfoliation (12) also called onion skin weathering, is the layered effect caused by expansion and contraction of the outer surface of hard rock which peels off these layers.
fault brecciation (14) is the grinding of rock material into small, angular fragments (breccia) within a fault as the two surfaces of the fault move past each other.
fjord (84) or fiord is the long, deep water course formed when a glacial valley enters the sea and is inundated with water.
flood plain (63) is the broad, flat surface on either side of the river on the coastal tract which is subject to inundation if river levees are broken.
fluvial erosion (54) is that caused by moving water in river systems.

free radical (23) is an uncharged molecule having an unpaired valency electron. They are typically highly reactive and short-lived.
frost heave (6) is the lifting of the rock and debris by ice as water freezes and expands.
frost wedging (13) is the splitting of rock as water in fine cracks expands as it freezes.
hanging valley (85) is a smaller u-shaped glacial valley which enters high up in the side of a larger glacial valley.
headwater tract (55) is the upper part of a river's course in the source area of the water.
helectites (49) are spelaeothems which grow in a variety of directions and as twisted shapes.
Hjulström diagram (53) is a graph of particle size against stream velocity showing the velocities required to erode, transport and deposit clasts within a river.
horns (87) or peaks are the sharp mountains formed by glacial action.
humus (32) is the dark, rich organic material forming the surface layer of some soils.
hydration (14, 27) is a form of weathering by which minerals such as clays, expand on the absorption of water.
hydraulic pressure (53) is the force of water upon the surface area of a rock face or soil.
hydrolysis (27) is the chemical breakdown of water molecules to hydrogen ions (H^+) and hydroxyl ions (OH^-), which can then react with minerals and be incorporated into crystal lattices.
illuviation (36) is the movement of mineral salts as solutions from one soil horizon to another by percolating water.
imbrication (68) is the aligning of flat pebbles on storm beaches or pebble bars in rivers such that their long axis faces downslope or downstream.
inselbergs (110) from the German for island mountain are large, isolated rounded rocky mountains which usually stand alone on desert plains.
isostasy (71) is the maintenance of equilibrium of the earth's surface often resulting in broad, gentle uplift or sinking.
karst (43) is a region in the northern Adriatic which gives its name to any exposed limestone area subject to massive weathering and erosion by water to produce sharp pinnacles, dolines (sinkholes)

and cave systems. Drainage of water in these regions usually is underground.

katabatic winds (95) from the Greek for flowing downhill are strong winds which blow from the interior of Antarctica down to the coast.

lacustrine (61) refers to the sedimentary environment of lakes.

lag deposits (60) are tear-dropped shaped gravel deposits with a bulbous end trailing off to a long tail downstream. They are often found in longitudinal bars and are formed by a sudden increase then decrease in the flow rate washing coarser gravels downstream during floods.

lahar (8) is a special form of fast-flowing mudflow formed when heavy rains or snow and ice melt on the unstable slopes of ash volcanoes.

landslides (8) are a sudden mass wasting of an entire slope as the surface becomes unstable and slips downhill.

leaching (21) is the removal of minerals from soils by moving water.

lee (91,104) refers to the side away from the wind, water or glacier.

levees (63) are raised river banks such that the bed of the river is often higher than the surrounding flood plain.

loess (31) from German for loose, this is very fine, wind-blown sediment derived from desert or glacial areas and carried long distances to fall and eventually accumulate as thick soils.

longitudinal bars (60) are long mounds of coarse gravel in the middle of long stretches of rivers deposited by floods.

longitudinal dunes (106) form where there is a constant prevailing winds and are usually parallel to the direction of the wind.

longshore current (67) is an ocean current which runs parallel to the coastline and is caused by the waves continually being forced around the headlands.

mass wasting (4) is the term used for the general removal of earth materials.

meander (57) is the bend of a river. It is not so pronounced in the steep headwater tract but on flatter flood plains it may become extremely broad and twisting.

mesas (109) are small desert plateaux, having a distinctive flat top and steep sides.

middle tract (58) is the area where the river slows down on a gentler gradient and begins to bend or meander more frequently.
moraine (86) refers to the piles of broken rock eroded by a glacier which form at the ends as terminal moraine, sides as lateral moraine and in the middle when two glaciers join as medial moraine. Often the terms moraine and till are used generally for glacial rock debris.
moulins (82) from the French meaning windmill, are round or curved holes in the top of a glacier formed as surface melt-water races around the hole like water going down a kitchen sink.
mudflows (8) occur when soil is enriched with water and becomes unstable, flowing rapidly downhill.
névé (82) is the young snowfield at the start of a glacier.
nitrogen cycle (21) is the recycling system nitrogen compounds of the earth especially of the nitrogen from the atmosphere into nitrogen compounds into the soil.
nunataks (94) are the isolated rocky peaks emerging from the vast ice covering of continental glaciers.
off-loading (12) occurs when deep-formed crystalline rocks such as granite crack into large, curved sheets as the rock is uplifted and the massive weight of the overlying rock is taken off by erosion.
organohalogen (23) compounds are those containing carbon, hydrogen, and oxygen with the halogen elements of chlorine, bromine and iodine.
overlapping spurs (57) as seen when looking down a young river valley, these are the ridges coming down from the highlands in an alternative orientation.
oxbow lake (60) which is called a billabong in Australia, is a curved depression, so-called because it looks like the old yoke put across the necks of oxen, usually filled with water which represents a meander that has been cut off from the main river which has changed its course.
oxidation (26) is the chemical reaction with elements to form oxides.
ozone (22) is the molecular form of oxygen gas which consists of three oxygen atoms O_3.
paludal (61) is the sedimentary environment of swamps.
parabolic dunes (106) form as vegetation stabilises one end of the dune, allowing the rest of the dune to migrate further.

pedestal rocks (103) are also called mushroom rocks or stone trees. They are unstable, mushroom-shaped rocks found in arid and semi-arid regions and have this shape because their base has been weathered down to a narrow column.
pedology (30) from Greek: *pedon* for soil and logos for study is the study of soils in their natural environment.
peneplain (64) is an extensive and uniformly flat area which has been eroded.
photosynthesis (22) is the process by which cyanobacteria (blue-green algae) and later plants, manufactured oxygen gas and sugars from atmospheric carbon dioxide and water.
piedmont plain (89) is a flat plain named from the Italian meaning at the foot of the mountains. They are often at the end of a glacial valley.
playas (111) from the Spanish *playa* for beach, develop as temporary lakes when water runs off the surrounding hills into large, closed desert depressions. These are often dry and contain sediment and evaporites.
point bars (60) are piles of gravels left on the outer bends of rivers.
reach (60) is a long, relatively straight stretch of a river.
regolith (30) is the general name for any broken rock material.
rheidity (82) is the ability for a medium such as water or ice to flow. Water may flow at different rates of centimetres or metres per second whereas ice may take several weeks to flow the same distance.
rimstone pools (50) or gours are walled pools, often found on the floor of a cave below an active dripping zone.
river terraces (58) are broad, flattened step-like structures formed on either side of a river where the river has continued to cut down through the built up river flats.
rôche moutonnée (91) from the French for sheep rock and also called whalebacks, are hump-like rocks, usually rounded at the top, sloping on one side and bulbous on the other formed as the glacier grinds over bedrock.
salars (111) are extensive salt lakes, often dry.
salination (41) is the impregnation of salt into a soil by solutions moving upwards through the soil or from other sources on the surface.

saltation (55) is the skipping motion of larger particles on the bed of streams caused by water flow or by smaller particles on desert platforms caused by the wind.

salt wedging (15) occurs on sea cliffs where saltwater from sea spray enters pore spaces or fine cracks of the rock face and the salt which crystallises as the water evaporates, pushes out softer particles of the rock.

sand spits (74) are long arms of sand deposited near the mouth of rivers **or** where there is an obstruction on the coast.

scree (7) is the loose fan-shape pile of angular rocks formed at the base of cliffs.

seracs (82) are the large blocks squeezed upward as a glacier moves over a hollow in the bedrock.

shawls (52) or curtains are a spelaeothem formed as a very thin ribbon of calcite, often only a few millimetres thick and translucent which grows out from the wall as water trickles down the surface.

slumping (5) is the general term used for the downslope slipping and folding of the soil surface.

soil creep (5) is the very slow movement of soil downhill.

soil horizon (34) is a distinctive layer of a soil profile.

soil profile (34) is a vertical cross-section of soil at a given place. It usually consists of several transitional layers or horizons.

solifluction (6) is a gradual movement of soil or rock fragments down a slope due to the constant freeze-thaw action which occurs in cold climates.

solution (28) is the dissolving or breakup of a solid in water forming ions or smaller molecules.

spelaeothems (47) are cave formations of crystalline material, usually calcite but sometimes gypsum if sulfate ions are present in the water, which grow as water of these solutions evaporates e.g. stalactites, stalagmites, helectites shawls, columns and rimstone pools (gours).

stalagmites (49) are spelaeothems which grow up from the floor.

stalctites (48) are spelaeothems which grow down from the ceiling.

star dunes (106) are pyramidal piles of sand with sharp tops which are formed when the wind blows from several directions.

storm beach (68) is formed when waves consistently hit the coast directly and deposit only coarser pebbles.

stoss (91,104) is the uphill side of a glacially-eroded rock on the side from which the glacier makes first contact.
striations (84) are the grooves and scratches left on bare rock surfaces by the movement of a glacier and the rock fragments it contains, over or past the rock face.
swale (76) is the long hollow between the frontal and hind dune of a beach dune system. It may be filled with water.
talus slopes (7) are steep slopes of broken, angular rock material called
thixotropic clays (11) which are generally hard and adhesive until shaken but which liquefy and collapse when shaken by earth tremors or man-made vibrations.
tombolos (74) or sand-tide islands are small islands joined to the mainland by a spit of sand which may be covered at high tide.
tors (17) are rounded boulders, often balancing on another, formed by the weathering of crystalline rocks such as granite.
till (79) such as usually glacial till, is the unsorted, irregularly-shaped pieces of rock formed by glacial erosion. It ranges in size from fine, rock flour to large, angular boulders.
tombolo (74) is the name given to small islands or sea stacks now tied to the mainland by a spit or bridge of sand. They are often called sand-tied islands.
tors (17) are large rounded boulders, often found balancing upon another, which are formed by the weathering of cracked blocks of hard, crystalline igneous rock exposed to the surface.
transverse dunes (106) are formed where there is abundant sand and the wind blows consistently from the same direction at ninety degrees to the direction of the dune.
truncated spurs (86) are the blunt-ended ridges cut off by a U-shaped glacial valley.
turbidites (11) are sedimentary rocks formed under the ocean by turbidity currents.
turbitity currents (11) are massive underwater avalanches of debris which suddenly break off the coastal continental shelf and flow down slope.
u-shaped valleys (84) are the rounded, curved valleys with almost vertical sides resembling a large letter u. they are carved out by the pushing grinding erosion of glaciers.
ventifacts (101) are any structure caused by wind erosion, such as the small, triangular rocks called dreikanters.

wadis (108) are dry water courses in desert regions which may be small gullies or very large valleys and broad depressions.
water cycle (3) or hydrological cycle is the great, natural system of recycling the earth's water.
water table (46) is the top level of water saturating a rock, soil or cave.
weathering (2) the breakdown of rock material by chemical and physical means where the rock has been exposed.
yardangs (103) are elongated and often bulbous rock structures eroded by the action of particle-laden wind in desert areas.
zeta-curved beaches (67) are the abruptly curved beaches resembling a hook or the Greek letter *zeta* when seen from above. They are produced when waves consistently strike the coast indirectly at an angle.

Books in the series **ADVENTURES in EARTH SCIENCE** are available in electronic format which can be purchased at https://www.amazon.com/ for Kindle or other electronic devices such as PCs and iPad using the free Kindle App. Books in the series are also available in print form (A5 – novel size) from Felix Publishing, Australia (info@felixpublishing.com) and include:

EXPLORATION SCIENCE
Field Geology & Mapping

RICHES from the EARTH
Minerals & Energy

ROCKS – BUILDING the EARTH

CHANGING the SURFACE
Erosion & Landscapes

FOSSILS – LIFE in the ROCKS
Ancient Lifeforms

A DANGEROUS PLANET
Volcanoes & Earthquakes

THROUGH SEA and SKY
Oceanography & Meteorology

BEYOND PLANET EARTH
An Introduction to Astronomy

Now released in electronic and print (A4) format is the complete reference book and text book suitable for Senior Secondary Schools, Junior College and Universities - **ADVENTURES IN EARTH SCIENCE.** Over 800 pages, 1200 illustrations (mostly taken by the author in many exotic places on seven continents), and over 30 video links on skills and virtual excursions to many parts of the world.

www.ingramcontent.com/pod-product-compliance
Lightning Source LLC
Chambersburg PA
CBHW070623300426
44113CB00010B/1629